Beyond Environmental Comfort

Beyond Environmental Comfort highlights some of the key ideas that form the foundation of the field of environmental comfort and, at the same time, gives voice to some of the concerns and considerations on the limitations of the field as it stands today.

Bringing together a range of foremost thinkers in their respective fields – Michel Cabanac, Derek Clements-Croome, Nick V. Baker, Sir Harold Marshall, Juhani Pallasmaa, Dean Hawkes, and Constance Classen – this book argues for a deeper appreciation of how environmental comfort may be understood in terms of our relationship with the environment rather than as independent qualities. For the first time these diverse views are brought together by Editor Boon Lay Ong to present insights into a world beyond what is normally covered in academic research. In the process, an attempt is made to define the field for the future.

This book shows that it is by understanding just how environmental design needs to go beyond mere comfort and deal with well-being that we can meaningfully design our future.

Boon Lay Ong currently teaches Environmentally Sustainable Design as course subjects and design studio at the University of Melbourne in Australia. He is also noted for developing the Green Plot Ratio, a sustainability metric that is used in Singapore to encourage greenery density.

Beyond Environmental Comfort

Edited by
Boon Lay Ong

Routledge
Taylor & Francis Group

LONDON AND NEW YORK

First published 2013
by Routledge
2 Park Square, Milton Park, Abingdon, Oxon OX14 4RN

Simultaneously published in the USA and Canada
by Routledge
711 Third Avenue, New York, NY 10017

Routledge is an imprint of the Taylor & Francis Group, an informa business

British Library Cataloguing in Publication Data
A catalogue record for this book is available from the British Library

Library of Congress Cataloging in Publication Data
Beyond environmental comfort / [edited by] Boon Lay Ong.
   pages cm
Includes index.
1. Architecture–Human factors.   2. Human comfort.   I. Ong, Boon Lay, editor of compilation.
NA2542.4.B49 2013
720'.47—dc23
                                        2012044321

ISBN: 978-0-415-45368-4 (hbk)
ISBN: 978-0-415-45369-1 (pbk)
ISBN: 978-0-203-11833-7 (ebk)

Typeset in Univers
by RefineCatch Limited, Bungay, Suffolk

Contents

List of Illustrations

Except where otherwise indicated, the illustrations belong to the authors of the respective chapters.

Chapter 1

Chapter 2

Chapter 3

Chapter 4

Chapter 5

Chapter 6

Chapter 7

Chapter 8

Chapter 9

Chapter 10

Chapter 11

Notes on Contributors

Nick V. Baker originally studied physics but has spent most of his professional life working in architecture and building, for much of his career at the Department of Architecture, Cambridge University. His expertise includes energy modelling, daylight design, natural ventilation and thermal comfort, as well as the broader field of sustainable design. He has written and contributed to several books on these topics, most recently *The Handbook of Sustainable Refurbishment: Non-Domestic Buildings* partly based on material from a recent EU-funded demonstration project, where he served on the expert panel. He is also co-author of *Energy and Environment in Architecture: A Technical Design Guide and Daylight Design of Buildings*, and *Climate Responsive Architecture: A Design Handbook for Energy Efficient Buildings*. More recently he has become interested in the way occupants interact with buildings, and in particular, the way their behaviour influences energy use. He is now retired but is still involved as a visiting lecturer and external examiner at several universities.

Michel Cabanac (MD) is a highly recognized physiologist who is influential in connecting hedonism (psychology) and thermal behaviour. While chairman of the IUPS Commission on Thermal Biology, he wrote the *Annual Review of Physiology* (1975). He also contributed to the *Handbook of Physiology* (1996) and to the *Encyclopaedia of Neurosciences* (1987, 1996, 2003). His deep influence on psychology can be recognized in his numerous invitations to address meetings of experimental psychology. Since his 1972 lead article in the journal *Science* on the physiological role of pleasure, he has pursued the experimental study of the hedonic dimension of consciousness, especially in decision making. This led him not only to new insights of consciousness itself and emotion, but also to a better understanding of human motivation in health as well as disease.

Constance Classen is the author of numerous essays and books on the cultural life of the senses, including *The Deepest Sense: A Cultural History of Touch* (University of Illinois Press, 2012); *The Color of Angels: Cosmology, Gender and the Aesthetic Imagination* (Routledge, 1998), *Worlds of Sense: Exploring the Senses in History and across Cultures* (Routledge, 1993), and *Aroma: The Cultural History of Smell* (Routledge, 1994, co-authored with

David Howes and Anthony Synnott). She is also the editor of *The Book of Touch* published in 2005 by Berg of Oxford. Recently she has been a visiting fellow at the Canadian Centre for Architecture where she investigated the social and sensory history of architectural design and museum display.

Emeritus Professor **Derek Clements-Croome** is founder of the MSc Intelligent Buildings course at the University of Reading and has carried out many research projects funded by the UK government. He is experienced in sustainable buildings research and education nationally and internationally and has sat on several UK research panels for EPSRC and ESRC. He chairs the CIBSE Natural Ventilation and Intelligent Buildings Groups and sits on the Schools Group. He is editor for the *Intelligent Buildings International Journal* published by Taylor and Francis. He has written many papers and written and edited several books including *Creating the Productive Workplace*, second edition, 2006 and *Intelligent Buildings*, 2004 (also in Chinese).

Dean Hawkes is Emeritus Professor of Architectural Design at Cardiff University and Emeritus Fellow of Darwin College at Cambridge University. He is currently Visiting Professor at the University of Huddersfield and De Montfort University, Leicester. In 2000 he received the international PLEA Award in recognition of his contribution to the teaching, research and practice of passive, low-energy architecture. In 2010 he received the biennial RIBA Annie Spink Award for Excellence in Architectural Education. His publications include *The Environmental Tradition*, 1996, *The Selective Environment*, with Jane McDonald and Koen Steemers, 2002, *The Environmental Imagination: Technics and Poetics of the Architectural Environment*, 2008 and *Architecture and Climate: An environmental history of British architecture, 1600–2000*, 2012. His built works have won four RIBA Architecture Awards, have been widely published in the professional press and exhibited at the Royal Academy of Arts Summer Exhibition in London and the Venice Biennale.

Emeritus Professor **Sir Harold Marshall** KNZM FRSNZ is an architect, engineer and physicist who is recognised internationally for his contribution to concert hall design. Formerly Professor of Architecture in the University of Auckland and Head of the Acoustics Research Centre, Sir Harold has more than 50 years experience in the acoustical design of auditoria and concert halls. His research with Michael Barron in the 1960s revealed the importance of lateral reflected sound in concert halls – a fact that has affected acoustical design profoundly since then. With Gottlob and Alrutz he first identified the necessary conditions for instrumental ensemble, and later extended this with Meyer to vocal ensembles. This study also produced the definitive measurements of the directivity of the singer's voice. In Marshall Day Acoustics he leads conceptual design. This is a role for which his architectural and musical skills have uniquely equipped him to communicate with architects and their clients. In 1994 he was elected a Fellow of the Royal Society of New Zealand. In 1995 he was awarded the Wallace Clement Sabine Medal by the Acoustical Society of America for his contributions to the field of architectural acoustics. In 2009, Emeritus Professor Marshall was knighted for services to acoustical science.

Boon Lay Ong is primarily interested in architecture, a diverting interest that lures him into many unfamiliar territories. His most recent publication is in thermal aesthetics, introducing not just the notion that thermal design in architecture has an aesthetic dimension but the argument that the thermal experience itself is aesthetic in nature. After completing his Bachelors of Architecture at Auckland University, he proceeded to a Master's in Acoustics. He later went on to study the role of plants (greenery) in architecture, leading to the development of a thermal computational model for plants, Green Plot Ratio and the notion of nurtured landscapes. At the National University of Singapore, he developed and was Director of the Master's of Landscape Architecture program and, later, was Director of the Master's of Science (Environmental Management) program. He is widely published and co-edited the book *Tropical Sustainable Architecture: Social and Environmental Dimensions*, with Philip Joo Hwa Bay. He is currently attached to the University of Melbourne, Australia, where he specializes in ecologically sustainable development.

Juhani Pallasmaa is an architect, SAFA, Hon. FAIA, Int. FRIBA and Professor of Architecture Emeritus. Pallasmaa has worked continuously as a designer, writer and educator since the 1960s. After leaving his position as Professor and Dean at the Helsinki University of Technology in 1997, he has held visiting professor positions in various universities internationally, most recently at the Catholic University of America in Washington DC. He publishes widely, mainly on the implications of human embodiment in art and architecture, and essays on individual artists and architects. His recent publications include *The Embodied Image* (London, 2011), *The Thinking Hand* (London, 2009), *The Eyes of the Skin* (London, 1995, 2005), *The Architecture of Image: Existential Space in Cinema* (Helsinki, 2001, 2005).

Preface

Boon Lay Ong

This book arose, as many of these books do, out of a casual conversation held over a cup of tea. On this occasion it happened when Professor Derek Clements-Croome was visiting the National University of Singapore where I was attached for the larger part of my academic career. I first got to know Derek through a paper I presented at a conference in Windsor, UK. He liked what I presented and asked me to write two chapters for his book, *Naturally Ventilated Buildings: Building for the Senses, the Economy and Society*. This time, he thought that I should edit a book to which he was willing to contribute if it suited my intended topic. Turn the tables, as it were.

It was a suggestion I welcomed. I am trained as an architect but my earlier school education had been almost entirely science based. While I may be good at the technical stuff, my heart belongs to the aesthetic end of things. It has been my great privilege in life to meet and be taught by many great experts in various technical fields who also understand the importance of the humanistic and the aesthetic in the work they do. It was a tremendous opportunity to put their work together in a single document.

The field of environmental comfort in architecture is an awkward one. Much of the research has been developed by engineers and scientists and hence tends to lack an adequate aesthetic or subjective understanding. This has resulted in a deep gap between research and practice. This gap is not much discussed in learned publications because the language to bridge this gap is currently underdeveloped.

In many ways, this book springs from that central disjoint between engineering and architecture. Engineers like their problems simple and precise while architects are always looking for the complexity in the design to work with. What the essays in this book show are that the two concerns do meet and where they overlap is where the most important part of environmental comfort is to be found. These areas of overlap are traditionally not written about. Dean Hawkes likes to remind me that some things are best left unsaid. Yet, I think there is a need to raise these issues and share these thoughts. They may be difficult to express well but they go to the very heart of what is excellent in the field. The authors gathered here are the best people to speak up about this.

This book sets out to do two things. It highlights some key ideas that form the foundation of the field of environmental comfort and, at the same time, it gives voice to some of the concerns and thoughts of various experts on the limitations of the field as it stands today. It looks both backwards and forward.

Chapters 1 and 2 set out the background for the field. As I explained in Chapter 1, environmental comfort is not quite officially a recognised field of research. It is subsumed under different headings and themes but its primary concerns are comfort in the fields of heat, light and sound, and with recent developments, the provision of Indoor Air Quality (IAQ). There is another field that combines these concerns into a composite quality called Indoor Environmental Quality. However, our concern here is not so much on defining indoor comfort but to argue for a deeper appreciation of how environmental comfort may be understood in terms of our relationship with the environment rather than as independent qualities. Michel Cabanac takes us into the laboratory in Chapter 2 to discuss comfort in biological terms. What Michel does is to link behaviour with stimulus, introducing a dynamic nuance to the provision of comfort. Environmental comfort is not simply the provision of conditions amenable for comfort but a result of both the environment and animal behaviour. We produce "an almost permanent thermo-neutral environment through clothing, shelter, heating and air conditioning". Later, he mentions food and water as well. For readers versed in the issues we are addressing here, Michel provides our first foray outside the field and introduces not only a more hedonic, but also more holistic, view. Although written in technical terms, he provides a launchpad for the views and arguments to follow.

Derek Clements-Croome takes this to the next level in Chapter 3, where he raises the relationship between consciousness and the senses and how they contribute to our well-being. In humans, environmental comfort is not just a physiological need, it serves our aesthetic and psychological appetites as well. These immeasurables impact our health and productivity – objective and quantifiable outcomes resulting from what may seem as somewhat hazy considerations. Where Michel has been technical, Derek is perhaps more philosophical.

The debate on adaptive thermal comfort, presented in Chapter 4 by Nick Baker, highlights the importance of the theme of this book. Nick argues that culture and genetics have been neglected in environmental comfort studies to our detriment. Modern architecture, in embracing environmental comfort as an entirely scientific endeavour, has been blindsided.

Though indoors, we are still "placed in the natural world and the building is seen only as a mediator". This contextual awareness has wide-ranging consequences. It is "reflected in a concern for the global environment – the choice of materials and a responsible attitude to the use of energy and other resources, messages which are implied by the design of the building".

Sir Harold Marshall reflects upon this through his acoustical practice and experience in Chapter 5. For Sir Harold, hearing is honest – revealing the material and volume of architecture through a sense that many of us do not consciously use. Acoustics is immediately and inextricably both subjective and objective at the same time. It is both aesthetic and scientific in Sir Harold's understanding – the one not quite achieved without the other. Echoing Michel's presentation of behaviour as response to the environment, Sir Harold reflects on various ways in which we adapt our behaviour to the acoustics of the environment – from prehistory, through the behaviour of children, students and an experiment with a young man profoundly deaf at birth amongst others. The auditoria for which he is consultant display an acoustical understanding that also results in visually breathtaking architecture. He rejects the engineering view that compromise between "architecture" and acoustical requirements is inevitable and accounts for the freshness of the designs by a meeting of minds between architect and acoustician. In turn that depends upon shared appreciation of meaning in architecture.

In his chapter on existential comfort, Juhani Pallasmaa discusses the hidden dimension of touch. Touch is understood in environmental comfort science mainly as heat, which is actually a subset of touch. As Juhani points out, touch is complex and our most primitive sense – of which the other senses are specializations and extensions. Touch brings out again the behavioural dimension of environmental comfort. Touch is what provides us with intimacy – "Through vision we touch the sun and the stars". Juhani further quotes, "Touch is the parent of our eyes, ears, nose, and mouth". Movement is touching with the feet and the body, experiencing time and space with this mother of the senses.

I follow in Chapter 7 with an observation about flight, in particular, about the effects of flying on our bodies. The vehicle is a unique approach to the design of an airport restroom. Restrooms (or toilets, to give them their more honest name) in airports are more than just a reprieve from the toils of travelling by air. They are our first introduction to our destination. As such, I proposed that we simulate the external environment in the airport restroom so that we may, through the act of refreshing ourselves, also be prepared for the environment of the place we are visiting. This is achieved using local greenery, the adoption of which requiring in turn an indoor environment that reflects the biorhythms of the place itself.

Dean Hawkes, in Chapter 8, introduces us to a novel approach to the appreciation of architecture. Aesthetics, in this chapter, is captured in the poetic sensorial experience of the spaces as imagined and realized by the architect. Though largely visual, the images captured in this chapter both in words and in pictures have thermal and acoustical dimensions that are as much a part of the aesthetic experience as the ocular aspect. Here, in these buildings, environmental comfort is broadened into environmental pleasure and satisfaction. The process of studying these buildings "involved mechanical recording through notes, sketches and photographs, but the principal tool was the actual experience of a building as time elapsed and qualities of light, air and sound, individually and in ever shifting combinations, were directly experienced. The only valid instrument was the human senses."

In Chapter 9, Derek makes another plea for the environment, developing his position further from Chapter 3. Though linked to his earlier chapter, much of what Derek now talks about has been given additional weight by the intervening chapters. There is a common voice, coming from different authors quite independently, sketching out an empathy for the environment that is particularly important and relevant at this time in our human civilization. Derek borrows the voice of other authorities, like Daniel Libeskind: "Buildings provide spaces for living, but are also de facto instruments, giving shape to the sound of the world. Music and architecture are related not only by metaphor, but also through concrete space." Derek also refers to the Sick Building Syndrome, thus touching upon indoor air quality, a topic neglected in the other chapters. Derek believes design affects our consciousness more deeply than many of us realize and concludes by quoting John Sorrell:

> We know that good design provides a host of benefits. The best designed schools encourage children to learn. The best designed hospitals help patients recover their health. Well-designed parks and town centres help to bring communities together . . .
>
> But true delight goes beyond the issue of beauty, it must also consider how the building contributes to the experience of those who use it, and whether it also makes a positive contribution to the community in which it is based.

In the penultimate chapter, I explore the environment as ecology. I embarked upon this research by reflecting on the simple fact that ecologists do not generally consider the built environment in ecological terms (except as a habitat for wild plants and animals that we tend to consider as pests) while architects and engineers are poorly skilled in ecology. The situation is much changed since I started on this research but there is much more still to be investigated. As Nick Baker noted in Chapter 4, "fifteen generations ago, a period of little consequence in evolutionary terms, most of our ancestors would spend the majority of their waking hours outdoors, and buildings would primarily provide only shelter and security during the hours of darkness". For a client, I designed a home based on ecological precedents – paradise, cave, clearing, and hydrology. I realized that themes of home and dwelling in architecture are themselves evolved from those places that we found and made home in wild and natural settings. Sustainable design is, to me, more properly ecological design and translates here, in the spirit that resides in all the chapters in this book, into form and space as well.

Constance Classen rounds out the essays here by postulating a sensorial urban future. She presents the aridity of current environmental studies well in her very first paragraph:

> At first glance "sustainability" and "pleasure" seem at odds. "Green" practices are commonly thought to involve an almost puritanical restriction of pleasures: shivering in frosty interiors to save on energy consumption, forgoing exotic foods in favor of homegrown staples, or walking weary miles to work rather than riding in comfort in a car. Surely "green" living describes an ascetic rather than aesthetic lifestyle. Beyond the satisfaction of feeling virtuous, what pleasures, what sensory enjoyments might living in a sustainable city offer?

Constance goes on to revisit the themes of this book by doing a critique of the modern city and speculating on what a sensorial city might be like. Constance points the finger of blame on the dominance of the visual in aesthetics – "*sonic, tactile, and olfactory qualities are ignored in contemporary urban and architectural designs, while visual effects such as monumental height or striking appearance are celebrated*". Because "*sight has functioned as the sense of domination, detachment, display, and cleanliness (in contrast to the more 'impure' sense of touch)*", our world is now anaesthetized and we are sensorially deprived. It is an ironic twist in our development, that the prominence of the visual in aesthetics would lead to the an-aestheticization of our environment.

Our sustainable future, as argued by the authors here, is not one simply concerned with the material resources of our society, the energetic demands of our consumption, and the detrimental impact of our industries on the natural environment. Our unsustainable present is as much unsustainable for us in social, psychological and aesthetic ways as it is in physical measurable ways. It is by understanding just how environmental design is concerned with more than mere comfort that we can meaningfully design our future.

The selection of authors in this book is perhaps fortuitous but they represent the foremost thinkers in their respective fields. The work here is deeply personal – a reflection of the kind of ideas in my head because of the teachers I have had. I had originally intended to collate a technical book, summarizing foremost research in this field. It is the authors who, knowing me, insisted on sharing the work that they know I most relate to. This is work you won't find them talking often about or publishing frequently, as in the academic world, much is made of technicalities. But it is work that we hold dear and talk about behind closed doors. Many of them I know personally – Derek Clements-Croome, Dean Hawkes, Nick Baker, Harold Marshall – while the rest I know by reputation and whose work I greatly admire – Michel Cabanac, Juhani Pallasmaa, Constance Classen.

Our backgrounds are rather diverse. Constance Classen publishes in sensorial studies and urban history. Michel Cabanac's work ranges from comparative physiology to thermoregulation while Derek Clements-Croome's interests cover both sustainable design and intelligent buildings, amongst other things. To him, the two terms *sustainable design* and *intelligent buildings* are interlinked. I met Dean Hawkes and Nick Baker at the Martin Centre at Cambridge University and they were both instrumental in getting my thoughts straight as I struggled with my PhD. Besides their academic careers, Dean is a highly successful architect and Nick is a physicist widely recognized for his work and insights in the field of thermal comfort. Having them both take an interest in my studies is like having my bread buttered on both sides. Sir Harold Marshall hails from New Zealand and was instrumental in my decision to go beyond my undergraduate studies and undertake research in the first place. I still remember, and repeatedly tell of, how he explained the importance of lateral reflections and how that inspired me to study acoustics. Academic research was not something that many aspiring architects considered useful to their careers at that time.

This book sits in the generally empty space between a fundamental textbook and a book on the state of the art. It assumes a general knowledge about the field of environmental comfort but does not go on to provide very much in terms of technical information about advanced research in the field. What it does do is to share some of the key thoughts that underpin the research directions that the field might take and provide insights into the experience and viewpoints that these experts have on the fields we are in. It hopes to encourage more architects to research and bring a broader view to what is often seen as a technical subject and to set out some signposts about what cross-disciplinary research in this field might be about. It also hopes to invite more engineers to engage with subjective or qualitative issues. After all, a number of the authors here are themselves engineers and scientists.

The main strength of this book lies in the link that the chapters here provide to other, perhaps more technical or specialized, material on this subject. Where the other materials do not provide, and are unable to because of their narrow focus, this book bridges, making available through the voices of the experts themselves where specialization and subjective experience come together. The reader will hopefully be inspired by the accessible nature of this book to delve deeper into other literature and come back again to reflect upon the links and other possibilities that the chapters here point to.

To this end, the style and layout of the chapters in the book are not homogenous. The various authors write in ways that best suit the material they want to discuss. The main concern has been to let the authors speak with as little encumbrance as possible. Key attributes of good academic writing have been preserved wherever possible. Most importantly, references where appropriate are identified and each chapter is followed with notes or bibliography. Some of the chapters have been published elsewhere but most of them are unique to this book. Where they have been published before, the original source is acknowledged in the first page of each chapter.

The book is written with this in mind – that we might explain, encourage, and inspire future research and design that will acknowledge the complementary roles of aesthetics and science in the making of architecture.

In closing, I'd like to thank the various authors who so generously contributed their time and expertise to this book. I can only hope that I have been successful in presenting their collective expertise and our shared passion to move environmental studies beyond the technical horizon. It is also appropriate that I mention again Emeritus Professor Clements-Croome who not only initiated this work but who has been unfailingly encouraging and helpful throughout. I seriously doubt if this book would have seen the light of day if it were not for him. It has been my great privilege to have been so mentored by him (as I have also been graciously enlightened in the presence of these luminaries).

At a more personal level, I am indebted to the leadership here at the Faculty of Architecture, Building and Planning, University of Melbourne. Two in particular stand out – Professor Tom Kvan, Dean, a leader with great foresight, charm and charisma and Professor Paul Walker, Deputy Dean, a man with much humour, patience and understanding. It is between them that I was given the space and time to work on this book. The various editors at Spon (later absorbed into Taylor & Francis), starting with Caroline Mallinder, then Georgina Johnson and finally Laura Williamson, have also been unbelievably kind and patient. My good friend, Raymon Ford, has also been an editor, in his case, unpaid. Last but not least, where would any man be without his precious and long-suffering wife, Fee Yoon and delightful daughter, Xin Hui.

Boon Lay Ong

Chapter 1

Introduction

Environmental Comfort and Beyond

Boon Lay Ong

At this time of writing, the field of *Environmental Comfort* does not in fact exist. This is so even though the topic is taught in nearly all schools of architecture and forms the core of much architectural discourse. It is to be found mainly in subjects like architectural science (e.g., Szokolay 2008),[1] environmental or climatic design (e.g., Drake 2009; Olgyay 1963), or architectural technology (e.g., Bougdah and Sharples 2009; Lechner 2009). The titles of these subjects allude to the scientific and engineering foundation of the topic and explain the emphasis on technology and scientifically derived criteria in current literature on the topic. To some extent, therein lie the constraints beyond which the current book attempts to expand.

As its name suggests, environmental comfort is about comfort criteria in the design of the built environment – in particular, in terms of *heat, light* and *sound*. Of more recent interest, environmental comfort today must now include *indoor air quality*. The first three topics emphasize the sensorial nature of environmental comfort and betray their origins in physics. Indoor air quality originated less from physics but from the discovery late in the twentieth century that modern buildings can cause illness, often as a result of air conditioning. Sometimes physical comfort and ergonomics may also be included amongst the topics. Since modern buildings are highly serviced, the subject includes the integration of engineering services like HVAC (heating, ventilation and air conditioning), lighting and acoustical systems. With current concerns on energy consumption and sustainability, the discussion is also steered towards how we can achieve comfort conditions with minimum energy and sustainable design.

This book, however, is focused on the fact that many researchers acknowledge the interconnectedness of the different senses and the limitations of scientific and laboratory-based criteria in capturing our holistic experience of architectural space. Lacking such holistic criteria, individualized criteria for each of these areas have found their way into building standards and are used by designers to ensure at least a regulatory level of comfort and satisfaction. In this book, key researchers in the various fields talk about their work as it pertains to extending our current understanding of the field.

In academic and scientific studies and research, the *environment* often refers to the large scale, as described in ecology, engineering and science. *Comfort*, on the other hand, is intimate, referring to the human body and its sensations. *Environmental comfort* lies in that middle space between our selves or bodies and the environment beyond. Besides the fields of architecture, building and engineering, the content of this book is also, in part, covered in nursing, hospitality and related fields, where the care of the human patient or customer is paramount. The emphasis there is different, as the latter practitioners have limited influence on the building and more interaction with the people in it. However, inasmuch as their activities are enacted within buildings, the concerns overlap and much is shared between these disciplines. What follows is a summary of the key concerns in environmental comfort that may be useful background knowledge to the other chapters in this book. Inevitably, much will be excluded in this summary and the material selected will seem inadequate. The reader will gain from perusing the many excellent textbooks and journals currently available for a wider and deeper understanding.

Environmental Comfort

Comfort is defined by Cabanac in the current book as "the state of physiological normality and of indifference towards the environment".[2] When successful, the objective of environmental comfort is thus reached when we are *indifferent* to our environment. However, *indifference* is possibly the furthest outcome from the architect's and even the client's objectives when conceptualizing the building and the spaces within.[3] Nevertheless, it is clearly necessary that our buildings should provide us with at least a level of physical comfort. After all, it is not the buildings themselves that are important but the fact that they facilitate or enable us to live and do various things within them.

Environmental comfort thus begins by setting out the criteria for comfort and traditionally separates into three primary concerns – heat, light and sound. All three sub-topics are approached by describing the sense itself and the attendant criteria necessary for comfort.

Our Senses

Heat, light and sound are not only sensory inputs but also forms of energy. Indeed, all our senses detect various forms of energy. Besides these three, we also detect mechanical energy with our bodies while our mouth and nose detect chemical energy through taste and smell. Four of the five organs that detect senses[4] are found on our faces while the fifth, the skin, covers our entire bodies both externally and internally.

Heat, light and sound are particularly significant in determining environmental comfort in buildings because they are the primary services that most buildings afford us. The technical considerations of many buildings will involve the provision of heating and cooling, of lighting and of reducing noise or enhancing desirable sounds. It is primarily to facilitate the design and specification of these services that comfort criteria are developed. Environmental comfort in all three fields varies depending on many factors including age, sex, physical fitness and cultural conditioning. With this in mind, environmental comfort criteria have evolved through statistical averages and will err on the side of convenience, social custom or economics. For example, it is easier and more socially acceptable to put on more clothing than it is to take them off. Thus, the design of most buildings will assume the normal level of dressing acceptable in that particular context and will accommodate the least amount of acceptable clothing in that situation. If cold, the individual can put on more clothing while it will be more socially difficult for the individual to wear less if warm.

Realizing that, on the one hand, we have our subjective senses, and on the other hand, objective energies that we are sensing, gives us a clue to the complexities that the authors here wrestle with. Our senses tend to act as a whole and relative to our needs. This leads to a given temperature being perceived sometimes as hot (after some strenuous exercise) or cold (if we have been sedentary for a long period before). It also leads to a blue room feeling quite cool and restful while a red room is often perceived as warm and energetic. The realm where these sensations are perceived, interpreted and understood is arguably the realm of aesthetics (see, for example, the chapters by Pallasmaa and Hawkes).

Heat

For a sense that is not actually mentioned as one of our five common senses (sight, smell, touch, taste and hearing), our ability to sense heat and cold is surprisingly important and surprisingly complex. It is, in fact, the most important of all the senses in the field of environmental comfort. A loss of this sense is more potentially lethal than the loss of any of the other senses. While there is precious little literature on environmental comfort, there is conversely an abundance of work on thermal comfort.

Our sense of thermal comfort is not just the sense of heat gain or loss between us and the environment. It is a reflection of the energy balance between our bodies and our environment. It is for this reason that our thermal sense is so important. If we lose more heat than our bodies can generate, or heat up more than our bodies can sustain, we run the risk of death. This thermal relationship also influences our level of activity. We are generally more alert and active when cool and slow down and become sleepy when warm. On long flights, some airlines may increase the temperature in the cabin to induce sleep. The physiology of heat involves not just the skin but our whole body. Our exhaled breath removes heat and water vapour as it expels air. When cold, we curl up, hunch our shoulders, rub our hands together and blow into our palms. We may even jump or hop on the spot to increase metabolic activity. When hot, we stretch and expand our bodies. As noted earlier, we tend to feel sleepy when warm and will generally slow down physical activity. The blood vessels at our extremities, the hands and feet, constrict in cold weather to help reduce heat loss. As a result, these extremities are often colder than the rest of our bodies when we are cold. Since our bodies are internally active even when we are physically inactive, we are most comfortable at temperatures approximately 10 to 15°C below our general body temperature. At a body temperature of 37°C, this translates to air temperatures between 20 and 25°C in most instances.

The primary way in which the thermal environment is described is in terms of *air temperature*. This is used both in referring to indoor thermal conditions as well as when we talk about climate. However, we gain and lose heat from the environment in other ways as well. *Direct radiation* from the sun is the most common way by which we warm ourselves. Solar radiation can also be reflected, as in the case of snow, or scattered, in the case of air molecules in the sky. The radiant heat from the sky is referred to in environmental science as indirect or *diffuse radiation*. In cities, external walls heated up by the sun also radiate a substantial amount of heat back to the environment. This has led to the phenomenon of the urban heat island effect or UHI where urban air temperatures are found to be more than 5°C higher than surrounding vegetated areas.[5] In fact, all objects radiate heat according to their surface temperature. Between objects of equal temperature, the amount of radiated heat given off and received are equal. Most of the rest of the heat we lose is lost through conduction to the air that touches our skins. As this air is warmed up, it rises and is replaced. This movement of air (or any fluid) caused by heating is called *convection*. The heat we lose is still conducted to the air on the skin but because the air is replaced through convection, we lose more heat than we otherwise might. When there is additional air movement caused by wind, we feel colder still. When there is wind, the effect of wind will replace the effect of local convection. As a result, the term convection is sometimes used in thermal calculations to refer to both convection and wind. More than other animals, we lose heat through sweat. Even when we don't feel it, our bodies sweat and keep our skin soft and wet. The presence of water vapour in the air influences the rate at which water on our skin evaporates. This, in turn, affects our sense of warmth. Humidity is thus a further factor in the heat equation.

The energy balance of our bodies can thus be defined as follows:

$$M = A + R + C + H$$

where   M = Metabolic rate, or internally generated heat
        A = Conduction (Air)
        R = Radiation
        C = Convection
        H = Humidity

The mathematical convention in the equation above refers to heat energy absorbed by the environment on the right and heat energy generated by the body on the left. Indoors, the radiant temperatures of the environment around us are generally at or around air temperature and are not felt as distinctively different heat sources. In fact, the radiant heat exchange is from us to the environment in most instances. However, very hot objects, like the sun or an oven, will emit sufficient radiant heat to be felt as a distinct heat source. To better represent actual conditions, the equation is usually expanded to include the effects of clothing. This is rather complex as it includes the kind, amount and coverage of the clothing we are wearing. In practice, equations like that in Figure 1.1, which predicts the theoretical mean vote based on laboratory studies, serve as the standard for thermal comfort in many countries.

While these four parameters (radiation, conduction, convection and humidity) jointly contribute to our sense of heat, we are able to distinguish between them. Thus we will describe our thermal environment not just in terms of hot and cold, but also in terms of wetness and windiness. The tropics, for example, are not only hot but also humid, making sweat a significant problem, while in temperate countries in summer the air temperature can be as high but because of low humidity, we feel more comfortable.

Our sense of thermal comfort is closely linked to climate. As a result, designing in response to climate and designing for thermal comfort are often discussed within the same work and synonymously. Olgyay (1963) reduced these considerations into a simple chart (Figure 1.2). What

$$PMV = (0.303e^{-0.036M}+0.028)\{(M - W) - 3.05 \times 10^{-3} \times [5733 - 6.99(M - W) - p_a]$$
$$-0.42 \times [(M - W) - 58.15] - 1.7 \times 10^{-5}M(5867 - p_a) - 0.0014M(34 - t_a)$$
$$-3.96 \times 10^{-8} f_{cl} \times [(t_{cl} + 273)^4 - (\bar{t}_r + 273)^4] - f_{cl}h_c(t_{cl} - t_a)\}$$

Where PMV = predicted mean vote, dimensionless
M = metabolic rate (46 to 232 W m^{-2})
W = external work (= 0 W m^{-2}, for most activities)
I_{cl} = thermal resistance of clothing (0 to 0.310 m^2 °C W^{-1})
f_{cl} = ratio of person's surface area when clothed to the area
       when nude, dimensionless
t_a = air temperature (10 to 30°C)
t_r = mean radiant temperature (10 to 30°C)
V_{ar} = relative air velocity (0 to 1 ms^{-1})
P_a = partial water vapour pressure (0 to 2700 Pa)
h_c = convective heat transfer coefficient, W °C m^{-2}
t_{cl} = surface temperature of clothing, °C

Figure 1.1 Fanger's PMV equation used in its entirety in ISO 7730 as the European and international standard for thermal comfort. Source: ISO 7730: 2005.

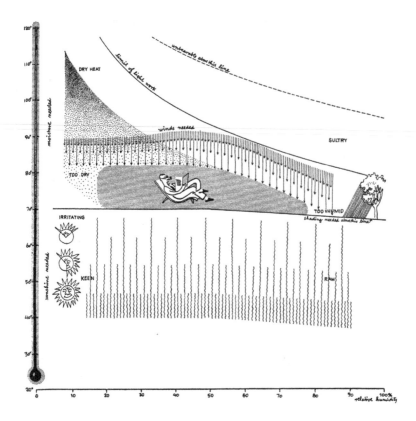

Figure 1.2 Thermal comfort conditions integrated into a single chart with temperature (in Fahrenheit) on the vertical axis and relative humidity on the horizontal axis. Source: Olgyay 1963, reprinted by permission of Princeton University Press.

the chart says is that indoor thermal comfort (no radiant heat, or sunlight, and no wind) is attained when the air temperature is between 70 to 90°F (≈21 to 32°C) within a relative humidity of 15 to 75 per cent. At temperatures below this, some sunlight (or other radiant heating) will be required and at temperatures above, some wind or air movement. Wind is needed for temperatures above 80°F (≈27°C) and this need increases with humidity. At humidities above 80 per cent, wind is needed at temperatures above 70°F. The number of lines for the sun and wind indicators suggests the strength of the sun or wind required.

Current thermal comfort criteria dispenses with any consideration of external climate. Through, most notably, the work by Ole Fanger and first published in his book, titled, appropriately enough, *Thermal Comfort* (Fanger 1970), current thermal comfort criteria only consider the immediate environmental conditions. If our thermal sense is simply a measure of our heat loss or gain from the environment, then what is important is the immediate thermal environment and not outside climatic conditions. In effect, the role of buildings is to protect us from the extremes of outside weather. In most modern offices, this immediate environment is artificially controlled and kept within a narrow range. This is reflected in the attire of office workers today, which remains largely the same in summer or in winter, in the tropics or in temperate countries. Other research carried out in the field argues differently. Nick Baker's chapter later in this book gives a hint of the debate. The significance of this debate is that if the *adaptive model* is accepted, then offices at least can be set at temperatures closer to the external.

Thermal comfort is of fundamental importance in the design of modern buildings. Temperatures beyond the comfort range are legally supportable reasons to stop work. Less than optimum temperatures affect productivity and health. In general, the comfort temperature is set around 23°C and an acceptable range falls within 20 to 26°C. This can be translated in practice as heating in winter up to 18°C and cooling in summer down to 30°C. Further adaption is assumed to be possible through more or less clothing. Beyond comfort and productivity, thermal comfort is also linked to heat stress. In Australia, where temperatures above 30°C are common, the Workcover Guidance to Working in Hot or Cold Environments (ACTU 2004) recommends rest breaks according to the following:[6]

- 30–32°C: ten minutes per hour;
- 32–35°C: 15 minutes per hour;
- 35–36°C: 30 minutes per hour;
- 37°C or higher: cease work until conditions improve.

Light

In literature on the senses and sensory aesthetics (e.g., Howes 2005), light and vision have predominated over the other senses. In the aesthetics of architecture (e.g., Winters 2007; Scruton 1992) the use of light to convey different moods and emotions is highly regarded and much discussed. In environmental comfort, however, the role of light is perhaps more mundane. The concern here is mainly to ensure visual clarity, defined in terms of adequate brightness, neutrality in colour and lack of glare. Access to daylight (or, at least, a sense of the outdoors) has been found to be important subjectively and helpful in giving inhabitants a sense of time.

Visible light is actually a narrow band of electromagnetic radiation – from about 400 nm to 700 nm. Placed side by side, this band is visually perceived as a band of rainbow colours (Figure 1.3). All the colours we can see can be reproduced by just three primary colours – red, green and blue. White is the result of the equal availability of all the visible light frequencies. If we include intensity, the human eye can distinguish approximately ten million different colours.

Our ability to distinguish colours is based on three types of colour receptor cells, or cones. Their responsiveness is not evenly spaced on the

Colour	Wavelength Interval	Frequency Interval
Red	700–635 nm	430–480 THz
Orange	635–590 nm	480–510 THz
Yellow	590–560 nm	510–540 THz
Green	560–490 nm	540–610 THz
Blue	490–450 nm	610–670 THz
Violet	450–400 nm	670–750 THz

Figure 1.3 Frequencies and wavelengths of the visible colour spectrum, or colours of the rainbow. The seventh colour, indigo, is often subjectively alike enough to violet to be considered a different shade of the same colour.

visible spectrum nor do they correspond strictly to the three primary colours. Instead, one type is relatively distinct from the other two and is most responsive to what we perceive as violet (~420 nm). They are often called S cones, or, misleadingly, blue cones. The other two types are closely related. The L cones, sometimes referred to as red cones, are actually most sensitive to yellowish green (~564 nm) while the M cones, or green cones, are most sensitive to light perceived as green, with wavelengths around 534 nm.

The final type of light-sensitive cell in the eye is called the rod, which is more sensitive than cone cells. These cells are most effective in dim light. There are approximately 20 times more rod cells than cone cells. The resultant dark image produced by cone cells is often colourless. Furthermore, the rods are barely sensitive to light in the "red" range. In certain conditions of intermediate illumination, the rod response and a weak cone response can result in colour discriminations not accounted for by cone responses alone.

Ambient light, often perceived as white or neutral, is often subtly shaded. Incandescent light tends to be yellowish while fluorescent light tends to be bluish. Natural daylight varies considerably over time – throughout the day, in different seasons – and also in different locations. This shading is generally identified in terms of *colour temperature*. As its name suggests, the colour temperature of a light source is the temperature of an ideal black-body radiator that has been heated until it radiates light of the given colour. Light with a low colour temperature tends to be yellowish while light with a high colour temperature tends to be bluish.

Comfortable Lighting

Lighting comfort or visual comfort is not as well defined as the field of thermal comfort. While specific research on visual comfort is available, there is no internationally recognized standard on visual comfort. Instead, there are standards on minimum lighting levels. The standards apply to ambient lighting, which is assumed to be neutral (whitish), diffuse and uniform. There are two key concerns: adequate lighting and glare. Adequate lighting is regulated in terms of minimum lighting levels for various tasks (Figure 1.4).

	Germany 1988	United Kingdom 1994	USA, Canada & Mexico 1993	Proposed European Guidelines
Standards	DIN5035	IES/CIBSE	IES(NA)	CEN/TC-169
Offices				
General	500	500	300	500
VDT Tasks	500	300–500	75	500
Desk	500	500		
Reading Tasks		300	300	500
Drafting	750	750	1500	750
Classrooms				
General	300	300	300	300
Chalkboards	500	500	750	500
Retail Stores				
Ambient	300	500–1000	300	300
Tasks/Till Areas	500	500–1000	300	500
Hospitals				
Common Areas	100		150	100
Patient Rooms	80–120	30–50	75	
Operating Room	1000	400–500	1500	1000–2000
Operating Table	20–100K	10–50K		20–100K
Manufacturing				
Fine knitting, sewing			1500-3000	500
Electronics	1000			1000

All lighting levels specified in lux.

Figure 1.4
Recommended lighting levels (lux) for selected countries. Most Asian and other developed countries adopt either the US, UK or European standard. Source: Mills and Borg 1999.

As in thermal comfort, human preferences differ with age, gender, external environment and other factors. Mills and Borg (1999) found that recommended lighting levels for various countries around the world rose from the 1930s to the early 1970s after which they tended to stabilize or decline. Not surprisingly, this trend matched the general trend of growing awareness of energy shortage and environmental crisis. Glare is somewhat more difficult to identify during the design stage. It is defined as a ratio between the object viewed and its surrounding lighting levels. Glare in lay terms also refers to specular reflections of the light source like when sunlight is reflected as a bright band across the wall or blackboard.

Lighting can be divided into the different functions it has to serve: ambient, task, accent and decorative. Many of the lighting standards in use are applied as ambient lighting. In fact, the recommended levels refer to task lighting. Ambient lighting need not be as bright and should in fact reflect external lighting conditions. While still inconclusive, recent research indicates that daylighting and an awareness of the external climatic conditions are helpful in terms of occupant health and well-being and may contribute to overall productivity (e.g., Aripin 2007). For general visibility, lighting levels of around 100 lux are usually acceptable. Bright lights affect human circadian rhythms and can even lead to the onset of certain cancers (Pauley 2004). Task lighting should be provided near task surfaces as fittings that users can manipulate to suit their needs. Accent lighting refers to lights used to highlight spaces or objects (e.g., paintings and sculptures) while decorative lighting are lights that are themselves meant to be aesthetically pleasing.

Sound

Sound is perhaps the distant cousin in the field of environmental comfort. Like heat and light, it is a complex field that is reduced into relatively simplistic standards to facilitate implementation. Noise, in particular, is of significant concern in the building industry. Too much noise is cause for quarrels between neighbours, legal action in courts, significant losses in property value and policy making in the community and urban planning. Nevertheless, under most circumstances, acoustics is not a major consideration in environmental comfort. This neglect of acoustical comfort is becoming untenable with the high noise levels in most cities today. While heat and light are generally important in most situations, acoustics become central in auditoria design like concert halls, theatres and cinemas.

The relationship between acoustics as a science and our subjective perception of sound is closer to its measurement[7] than it is for light and heat. The primary unit for measuring sound is decibels, defined as ten times the logarithmic ratio of the sound intensity relative to the threshold of human hearing (set at 20 micropascals, or μPa).

$$dB = 10 \ (P/P_0)$$

where   dB = Sound Power Level (SPL)
         P = Sound power of source
         P_0 = Reference sound power (20 μPa)

From Figure 1.5, it would seem that we are most sensitive to sounds around 2 to 4 kHz. Actually, this is amongst the highest frequencies of the sounds we normally hear. The piano, which has the widest range of notes amongst musical instruments, has a frequency range of around 30 Hz to 4 kHz only. The normal octave,[8] starting from what is known as Middle C, ranges from around 260 to 520 Hz[9] and is tuned around A4, set at 440 Hz. Most of us, if we could sing at all, would sing most comfortably around this range of frequencies. A study (Gilbert and Robb 1996) of baby cries within the first year of birth locates the fundamental frequencies of their cries within the range of 400–500 Hz with the pitch rising from birth to the end of the first year. We are least sensitive to low frequencies but because of the energy required in their production and their large wavelengths, they are the hardest to attenuate.

Both noise criteria and sound levels are specified and measured with reference to human hearing. Noise levels are generally discussed in terms of dB(A), which is sound weighted in a curve that simulates human hearing. Other weightings exist but are rarely used. Most acoustical professionals

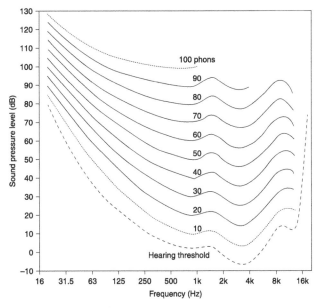

Figure 1.5 Equal loudness contours determined by recent research. Source: Suzuki and Takeshima 2004.

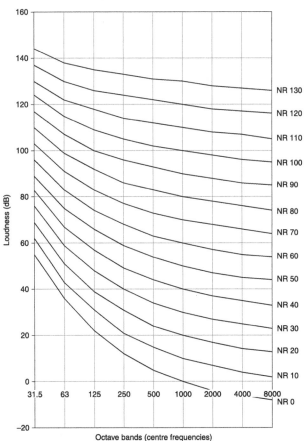

Figure 1.6 Noise Rating Curves are used to specify acceptable background noise levels for acoustical design. Notice the similarity and imprecision between these NR curves and the equal loudness contours. The NR curves are specified according to their dB levels at 1 KHz. Data retrieved from ISO 1996–1:2003.

would instead do a more complete analysis of the sound and measure it in octave or third-octave bands. Internationally standardized octave bands are designated by their centre frequencies starting with 31.5 Hz.

One of the consequences of the logarithmic properties of sound is that high absorption values do not compute into high attenuation. An absorption of 90 per cent of the sound results only in a 10 dB drop in noise level; 99 per cent absorption will only improve this to 20 dB. A typical timber frame wall, which gives an attenuation of around 35 dB, amounts to a 99.97 per cent reduction in sound intensity. Thus, sound-absorbent materials used to reduce the reflections in a room are effective only in reducing the noise level within the room and do not substantially affect the noise level on the other side of the room.

Indoor Air Quality

Indoor air quality (IAQ) became a global issue in the 1980s with the spread of concern over the sick building syndrome or SBS (WHO 1984). Sundell (2004) points out that historically there has been concern over polluted and damp air at least since Greek and Roman times. Sick building syndrome is a particular issue with modern air-conditioned buildings arising from the high volume of recycled air leading to increased concentrations of CO_2, volatile organic compounds (VOC), asbestos, bacteria, dust, tobacco smoke and other pollutants. The low temperatures of air conditioning also result in high humidity and condensation, which encourage mould growth.

Thermal comfort, as discussed earlier, can affect occupant health and productivity and is sometimes regarded as part of indoor air quality (sometimes broadened and referred to then as indoor environmental quality or IEQ). The effects of poor IAQ can range from just general tiredness to migraines and allergic reactions that incapacitate the worker. The response to poor IAQ depends a great deal on the individual, his physiological sensitivities, state of health and stress levels. SBS is hard to isolate as these symptoms can also be linked to allergies and illnesses unrelated to the building.

The most immediate and successful way to improve indoor air quality is to increase natural ventilation (WHO 2009 and 2010). This can be problematic if the outside air is too cold or warm or polluted. It can also be a problem if opening the windows exposes the occupant to external noise as well as fumes. Plants (or greenery) emit VOCs as well but normally not in high enough concentrations to be a problem. On the other hand, airborne pollen from plants can bring on hay fever and asthma attacks.

As well as ventilation, the careful selection of materials in terms of finishes and furnishings is also important and many manufacturers today will indicate if their material is VOC free or safe in terms of volatile substances. Finally, care during demolition, construction and renovation is essential to prevent or reduce contamination to the environment (particularly the neighbours) due to building processes.

Climate and Environment

Buildings moderate our relationship with the external environment. They keep the elements (sun, wind, rain and snow) out and, when desired, enable us to control them to within desirable parameters. For the most part, this translates into a focus on the thermal environment. Our sense of heat corresponds well to the main elements of climate – sun (radiation), air (temperature), wind (convection) and rain (humidity). It is indeed our sense of the weather. Modern architecture, with access to compelling technology, can and has generally ignored the external environment and met human comfort needs through engineering (Banham 1969). Traditional cultures are not quite so "fortunate" and have evolved more benign ways to mediate between human needs and the external environment (Olgyay 1963; Dahl 2010; Oliver 2003). As a result, traditional architecture is much more directly shaped and informed by the climate within which it resides. Even the clothes we wear, the food we eat, the times at which we work or sleep are deeply influenced by climate. In being able to artificially overcome climatic constraints, we have been able not just to produce any architectural form or style anywhere we like, we have also been able to live, work and play more freely as we please.

This freedom comes at a huge energy cost and the challenge of sustainability is to continue to enable such freedom while reducing the excesses that past and current practices have engendered. Proponents of modern architecture have not been unaware of the importance of climatic response and the energy demands of their buildings. Several, if not all, of the books referenced in this chapter attest to this concern. Despite this, due perhaps to a combination of cheap energy and globalization, architects, engineers and users alike have tended to trade energy consumption for greater flexibility, convenience and pleasure.

Beyond Environmental Comfort

Current concerns about sustainability have renewed research on environmental comfort. Much of this research may be found at the technological end in terms of increasing the efficiency of engineering services and finding alternative sources of energy. However, the fundamental criteria for environmental comfort have also come into question. When the cost of and pollution from energy are not of concern, standards have tended to overprovide – to be cooler or brighter than necessary.

What research is beginning to highlight is that environmental comfort cannot be understood independently of our subjective relationship with the environment. With thousands of years of civilization behind us, we have adapted to and developed different ways to live in harmony with the environment in different locations. These adaptations and ways have made lives in different places richer and more diverse as we learn to work at different paces, eat and sleep at different times and value the environment differently. The papers here collectively argue that we need to reach beyond current environmental standards to engage our subjectivity and, in the process, find that we are more adaptable and can even take delight in a wider range of environmental conditions than is acceptable within current comfort standards. The key denominator seems to be an engagement with the environment beyond rather than a narrow focus on immediate surroundings. The discourse around thermal comfort is telling in this regard. The dispute, if we may call it that, is between the scientific rigour of laboratory research and the wider range of comfort temperatures found through field studies. At its heart, it is not a scientific dispute but one that argues for a greater recognition of the diversity of human culture. We need to allow people in different places to work (or not work) according to their needs and the demands of the environment, appreciating the fact that different ways to be productive can be found to meet different circumstances. In the process, the buildings we build will not be anonymous and uniform regardless of their location but rich in character and experience. What lies beyond environmental comfort is green pleasures.

Notes

1. While the references are necessarily current, it should be noted that the fields are quite well established and several of the books are updated versions of books first published several decades earlier.
2. Chapter 2.
3. See Hawkes, for example, later in this book, for the kind of effect that architects are dedicated to pursue in their designs.
4. Most studies on the senses acknowledge more than the five common senses and will include the senses of balance, direction and movement among the scientifically accepted senses. The senses of heat and pain are somewhat contentious, in some cases included as the sense of touch, in others, treated separately. The four sense organs on our faces are the ears, eyes, nose and tongue.
5. The Earth Observatory at NASA provides good up-to-date information on UHI, e.g. http://earthobservatory.nasa.gov/IOTD/view.php?id=47704 (accessed 29 January 2013).
6. *Thermal Comfort Guidelines*, www.utas.edu.au/__data/assets/pdf_file/. . ./Thermal-Comfort-Guidelines.pdf (accessed 19 February 2011).
7. It is so close that Sir Harold Marshall calls it the honest sense in his chapter in this book.
8. An octave is a curious acoustical term. It refers to a range of frequencies in which the highest frequency is twice that of the lowest. This relationship results in a subjective hearing of the two sounds as being similar but different only in pitch. The term *octave* is derived from the musical scales where eight notes form an octave.
9. Actually 261.626 Hz at Middle C and 523.251 Hz an octave above.

References

Aripin, S. (2007) 'Healing architecture': daylight in hospital design. Conference on Sustainable Building South East Asia, 5–7 November, Malaysia.

Banham, R. (1969) *The architecture of the well-tempered environment*. London: Architectural Press.

Bougdah, H. and Sharples, S. (2009) *Environment, technology, and sustainability*. Abingdon: Taylor & Francis.

Dahl, T. (ed.) (2010) *Climate and architecture*. Abingdon: Routledge.

Drake, S. (2009) *The elements of architecture: principles of environmental performance in buildings*. London: Earthscan.

Fanger, P. O. (1970) *Thermal comfort: analysis and applications in environmental engineering*. New York: McGraw-Hill.

Gilbert, H. R. and Robb, M. P. (1996) Vocal fundamental frequency characteristics of infant hunger cries: birth to 12 months. *International Journal of Pediatric Otorhinolaryngology*, 34(3): 237–243.

Howes, D. (ed.) (2005) *Empire of the senses: the sensual culture reader*. Oxford: Berg.

ISO 7730:2005, Ergonomics of the thermal environment — Analytical determination and interpretation of thermal comfort using calculation of the PMV and PPD indices and local thermal comfort criteria. www.iso.org/iso/catalogue_detail.htm?csnumber=39155 (accessed 19 February 2011).

Lechner, N. (2009) *Heating, cooling, lighting: sustainable design methods for architects*. Hoboken, NJ: John Wiley & Sons.

Mills, E. and Borg, N. (1999) Trends in recommended lighting levels: an international comparison. *Journal of the Illuminating Engineering Society*, 28: 155–163.

Olgyay, V. (1963) *Design with climate*. Princeton, NJ: Princeton University Press, renewed 1991.

Oliver, P. (2003) *Dwellings: the vernacular house world wide*. London: Phaidon.

Pauley, S. M. (2004) Lighting for the human circadian clock: recent research indicates that lighting has become a public health issue. *Medical Hypotheses* 63: 588–596.

Scruton, R. (1992) *The aesthetics of architecture*. Princeton, NJ: Princeton University Press.

Sundell, J. (2004) On the history of indoor air quality and health. *Indoor Air*, 14(7): 51–58.

Suzuki, Y. and Takeshima, H. (2004) Equal-loudness-level contours for pure tones. *Journal of the Acoustical Society of America*, 116(2): 918–933.

Szokolay, S. K. (2008) *Introduction to architectural science: the basis of sustainable design* (second edition). London: Architectural Press.

WHO (1984) *Indoor Air Quality Research*. Copenhagen: World Health Organization Regional Office for Europe.

WHO (2009) *WHO guidelines for indoor air quality: dampness and mould*. Copenhagen: World Health Organization Regional Office for Europe.

WHO (2010) *WHO guidelines for indoor air quality: selected pollutants*. Copenhagen: World Health Organization Regional Office for Europe.

Winters, E. (2007) *Aesthetics and architecture*. London: Continuum.

Chapter 2

Sensory Pleasure and Homeostasis

Michel Cabanac

Since the origin of life, animals have behaved so as to seek environments favorable for their physiology and survival. All basic physiological needs are met by behaviors, and, of course, the first stage of sexual reproduction is also behavioral. Behavior is therefore the first and most powerful response by an animal to achieve its physiological aims and maximize its chance of survival.

The adaptation of behavior to the physiological needs is made possible because the central nervous system receives useful information on the state of the environment as well as that of the body itself (Figure 2.1). The present chapter deals with the way these signals participate in the various physiological regulations that maintain a stable *milieu intérieur* through behavior. A deliberate mentalistic attitude is adopted in the following. Such an approach proved fruitful for the understanding of the role of pleasure in the case of decision making when non-physiological motivations enter into play (Cabanac, 1992). The present chapter, however, concentrates only on sensory pleasure. The mechanisms described will be mostly short term, but the long term will also be envisaged.

Physiology and Behavior

From the time of the Greek philosophers for whom behavior was biology, there has been a continuous tradition for consideration of behavior as a part of physiology. The first students of what we now consider to be the

Figure 2.1 Psychologists and Ethologists study the various behaviors as functions of the signals received by the central nervous system from the environment and the body itself. This "behaviorist" attitude considers the brain as a black box. Yet in the black box a mental phenomenon (motivation) takes place and can be correlated to the various physiological behaviors. The study of this mental phenomenon is the "mentalist" attitude.

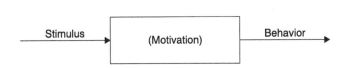

science of physiology were, in fact, concerned with behavior, motivation, and cognition.

Thus, Johannes Müller (Müller, 1834–1840) studied sensations, while Claude Bernard (Bernard, 1855) studied the determinants of thirst and its behavioral response, water intake. Pavlov (Pavlov, 1966) discovered conditioning and conditioned reflexes as a development of his studies on digestive secretions. This trend was also followed by Cannon (Cannon, 1932) who studied hunger and emotion in humans and animals, and by Hess (Hess, 1936) who was the first to investigate coordinated behavioral responses after either electrical stimulation or lesion in the diencephalon of behaving animals. Although the study of behavior as an integrated physiological response has persisted during the last 50 years, behavior has been a minor branch of physiology. This reductionist trend has also occurred among neuro-physiologists who tend to study the functional anatomy of the brain from the point of view of treatment of signals. Behavior is used by them mostly as a convenient window through which to probe the structure of the brain. In the past, a good reason for this development of neurophysiology may well have been the constraints imposed by the use of anesthetized animals and of stereotaxic apparatus that did not permit animals to behave. Meanwhile, behavior was studied by natural scientists and by psychologists.

The entry of behavior into natural science is usually attributed to Darwin (Darwin, 1872). Later, behavior was studied in Europe by zoologists who founded the science of ethology, which is the scientific study of the behavior of animals integrated in their natural environment. In 1951, Tinbergen introduced the term ethophysiology (Tinbergen, 1951). Ethology is largely devoted to the study of animal fitness, which includes not only reproductive success, but also how behavior optimizes physiological func-tion, i.e. whole-body physiology. An important branch of ethology is behav-ioral ecology (Krebs and Davies, 1981) where the integration of animals into their environment is studied and measured quantitatively.

Until the end of the eighteenth century psychology was the branch of philosophy that studied the human mind. Psychology became a science, i.e. experimental psychology, only when the study of behavior became experimental, under the influences of psychophysicists such as Helmoltz (Helmoltz, 1879) and psychologists, such as Thorndike (Thorndike, 1898) and Watson (Watson, 1919). Psychology started to diversify; one important branch of that science is physiological psychology in which behavior is studied for itself or as a means of exploring motivation. As a result, behaviors with clear physiological aims, such as food and water intake, have been studied mostly by psychologists (Richter, 1936; Stellar, 1954).

How the body satisfies its needs is left mostly, therefore, to psychologists and how it quantitatively adjusts to the environment is left mostly to ethologists. Traditional physiology deals with the functioning of the various systems within the body. The modern trend is toward a finer and finer level of analysis of the systems whose sum constitutes the living organism. Modern physiology has become so analytical that its borderline with biochemistry hardly exists any more. Thus, physiologists may be losing sight of an aspect of paramount importance: The living being is not a closed system, isolated from the environment, but is open and exchanges energy and matter with the environment. Physiologists take too often for granted the permanent inflow from the environment into the body, and outflow from the body to the environment, of whatever variable they are concerned with.

The return of behavior to departments of physiology is recent and has occurred largely under the influence of environmental physiologists who are concerned with the adaptation of living organisms to their various and variable environments. Environmental physiology, by its very nature therefore, is integrative and deals in most cases with the functioning of the entire organism. Environmental physiologists, of course, have always been aware of the environmental constraints that influence life processes, especially in the case of energy balance. Early studies showed that behavior was the only response used by reptiles to maintain a stable core temperature (Langlois, 1902). However, it is only recently that the majority of physiologists have acknowledged that behavior is a powerful way to cope with environmental constraints. Temperature regulation is mostly behavioral and was therefore important in heightening the awareness of physiologists so that they should not ignore behavior. Hardy (Hardy, 1971) was among the first contemporary physiologists to consider behavior as deserving to be studied, and Bligh (Bligh, 1973) pointed out that one should not oppose "physiological" and "behavioral" temperature regulation; since the behavioral response is also physiological, he suggested the use of the terms "autonomic" and "behavioral" temperature regulations.

At present, the concept that behavior belongs as much to physiology as to psychology and zoology (ethology) is not yet commonplace, but there are signs that the pendulum is swinging back and that behavior again is a subject of challenge for curiosity-minded and thoughtful physiologists. A major contribution in that direction was the founding of the journal *Physiology and Behavior* by Matt Wayner in 1966.

Behavior, from Simple to Complex

Simple taxis and migration to better micro-climates were the only behaviors of unicellular animals. With an evolutionary increase in size and in physical complexity, other behaviors were to emerge. First came simple postural adjustments, then parental behaviors and micro-environment building. The ultimate effectiveness of behavior is achieved in the human species. Humans live permanently in artificial environments, protected by their clothes and dwellings. As a result, the autonomic responses of shivering and sweating for temperature regulation are almost never seen in ordinary life because human behavior produces an almost permanent thermo-neutral environment through clothing, shelter, heating and air conditioning. Similarly, an internal water balance is permanently achieved with drinking – a behavior – although a limited amount of water is produced metabolically. The needs for energy and the body's building blocks are met by behaviors: Eating, food hoarding, and food storage. A final physiological need, respiration, is fulfilled mostly by an autonomic response, i.e. ventilation, but may occasionally initiate behaviors for survival in extreme environments such as at high altitudes or under water.

All this seems obvious to students of human behavior, not only to the public but also to many scientists. Yet few question the nature of the link between physiology and behavior. Most people are satisfied with the word "motivation." Food intake, or temperature regulation, and other behaviors are "motivated"; that is explanation enough. Such a semantic step, though necessary, is not sufficient. What is motivation? To answer the question one must acknowledge the existence of consciousness, an invisible property emerging from the complexity of our brain. In the following, I shall defend the hypothesis that pleasure is the dimension of consciousness that motivates those behaviors that increase our chance of survival; and, especially, that the trend to maximize sensory pleasure leads to behaviors that optimize the physiological state of our body. Further, I shall offer the hypothesis that sensory pleasure emerged in the vertebrates ancestral to present-day reptiles, birds, and mammals.

The information that is used to adjust behavior to the body's needs comes from sensors and produces sensations. Sensation is a four-dimensional object of consciousness (see Cabanac, 1996). The dimension that motivates behavior is hedonicity.

Human Sensory Pleasures

When a stimulus excites a sensory neurone, it arouses a four-dimensional sensation: Quality, intensity, pleasure/displeasure and duration. The qualitative dimension identifies the nature of the stimulus: Not only visual, auditory, etc., but also the color, shape, melody, etc., of the stimulus. The intensive dimension describes the intensity of the stimulus: Brightness of a light, strength of a sound, concentration of an odor, etc. The hedonic dimension that can be absent describes the usefulness of the stimulus, as will be developed below; pleasure is related to usefulness, displeasure to danger or harm and indifference to uselessness. The fourth dimension is time and describes the duration of presentation of a stimulus.

The hedonic part of a sensation is the amount of pleasure or displeasure aroused by the stimulus. According to Young (Young, 1959), this dimension of sensation is a continuum from extreme negative hedonicity (distress) to extreme positive hedonicity (delight), with indifference in the middle. Sensory pleasure possesses several characteristics: Pleasure is contingent, pleasure is the sign of a useful stimulus, pleasure is transient, pleasure motivates behavior.

Sensory pleasure is not fixed but contingent and depends on (i) the nature of the stimulus, (ii) the internal state of the subject, and (iii) the past history of the subject. A given stimulus can arouse pleasure or displeasure according to the combination of these three parameters.

Not all stimuli evoke pleasure or displeasure. In the vast, permanent flux inputs from the sensors to the central nervous system, the large majority elicits an indifferent sensation. For example, the sight of most objects is neither pleasurable nor displeasurable. Only a small minority of the innumerable visual, auditory, and even olfactory stimuli that reach our senses at each instant evokes pleasure or displeasure. Most visual objects, sounds, and odors evoke neither pleasure nor displeasure. The sight of a chair in my office is most of the time indifferent to me. The sound of students speaking in the corridor is usually indifferent, etc. However, the smell of a meal is pleasurable at lunch time, the cold of an outside environment is unpleasant in winter. In addition, as stated by Pfaffmann (Pfaffmann, 1960), "there is almost no limit to the range of previously neutral stimuli that, by one method or another, can be made pleasurable or unpleasurable." Pfaffmann designated as primary reinforcers all stimuli creating sensations of pleasure or displeasure. The majority of them are negative reinforcers. A stimulus can be unpleasant by its very nature (e.g., a bitter taste) or by its intensity (e.g., a violent sound). When the intensity of a neutral stimulus increases, the sensation usually becomes unpleasant, as Wundt (1874) proposed. Thus, pleasant stimuli are a minority, and it is striking to observe that the range of pleasantness is limited both qualitatively and quantitatively. The most effective, and probably only, primary positive reinforcers are chemical, thermal, and mechanical stimuli.

Alliesthesia is the faculty of a sensation to move up and down the hedonic axis of Figure 2.2. The word alliesthesia (Greek *Aliosis* – changed and *esthesia* – sensation) is applied to the hedonic component of sensation, pleasure, or displeasure. The amount of pleasure or displeasure aroused by a given stimulus is not invariable, it depends on the internal state of the stimulated subject and on information stored in memory. Factors that can modify the internal state and in turn induce alliesthesia are as follows: Internal physiological variables (e.g., deep body temperature or body dehydration modify the pleasure of thermal sensation or taste of water); set-points (e.g., during fever the body temperature set-point is raised and pleasure defends the elevated set-point); multiple peripheral stimuli (e.g., mean skin temperature determines the set-point for deep body temperature and in turn generates alliesthesia); and past history of the subject (e.g., association of a flavor with a disease or a recovery from disease renders it unpleasant or pleasant). Positive alliesthesia indicates a change to a more pleasurable sensation, negative alliesthesia a change to a less pleasurable one. For example, the taste of a candy is pleasant if I am hungry; if I eat ten candies in a row, the eleventh may arouse displeasure: Negative alliesthesia took place with this sweet sensation (e.g., Figure 2.3). If I try again to eat a candy a couple of hours later, now I may feel pleasure: Positive alliesthesia occurred since the last ingestion. This will be described and studied more systematically in the following pages.

Sensory pleasure describes the usefulness of the stimulus. Useful stimuli arouse pleasure and noxious stimuli arouse displeasure. Usefulness is understood here as improving physiological fitness, i.e., as the capacity for the stimulus to correct a physiological trouble or deficit. Usefulness of sensation is thus appreciated from its short-term survival value, but learning can extend to the long-term survival value the association of sensory pleasure with long-term usefulness.

Sensory pleasure is transient. Since the usefulness of a stimulus can change with time, so does its pleasantness. In the short term, sensory pleasure is only transient, lasting only as long as the physiological state has returned to normal. Then the stimulus arouses only indifference or displeasure. Different from pleasure is comfort: The state of physiological normality and of indifference towards the environment. Comfort is the state of sensation with a nil hedonic dimension. Whereas sensory pleasure is dynamic, aroused when a stimulus is useful, comfort is a stable state that can last indefinitely if the environment and the subject remain in stable conditions.

Sensory pleasure is the motivation for physiology oriented behavior. There is an obvious relation between the hedonic part of sensation and behavior; the strength of the motivation for or against a stimulus is a function of the intensity of the pleasure or displeasure elicited by the stimulus. Pleasant sensations induce approach or consummatory behaviors (or both) for alimentary, sexual, and thermal stimuli. Relations exist between pleasure and usefulness and between displeasure and harm or danger. For example, the sweet taste, which arouses pleasure in fasted subjects, is associated to energy-rich molecules that will contribute to cover the needs of energy metabolism. On the other hand, a bitter taste is often associated to molecules with pharmacologic actions that might be harmful or even fatal. Thus, the evidence supports the hypothesis that sensory pleasure is a sign of usefulness and displeasure a warning sign. The pleasure or displeasure aroused by a thermal stimulus can be predicted from the various body temperatures of the person stimulated. Pleasure is actually observable only in transition, when the stimulus aids the return of the subject to normothermia, all stimuli lose their strong pleasure component and tend to be indifferent or unpleasant. This scarcity of pleasure may be more apparent than real because temperature regulation is never achieved, and normothermia is an almost virtual situation. Sensory pleasure and displeasure thus appear especially well suited as motivation for thermoregulatory behavior (Attia, 1984). The case of pleasurable flavors shows an identical pattern. Alimentary flavors are pleasurable during hunger and become unpleasant or indifferent during satiety. Measurement of human ingestive behavior confirms this relationship (Fantino, 1984). Preference shows a qualitative influence: Human subjects ingest more of what they like. It also shows a quantitative influence: The amount eaten is a function of the alimentary restrictions and increases after dieting, when negative alliesthesia has disappeared. Thus, the seeking of pleasure and avoidance of displeasure lead to homeostatic behaviors. Sensory pleasure therefore indicates a useful stimulus and simultaneously motivates the subject to use it. Both a reward and a motivation, pleasure leads to optimization of life.

When placed in a conflict of biological motivations, e.g., fatigue versus thermal comfort, human subjects tend to maximize the sum of their sensory pleasures (Cabanac, 1992). By so doing they optimize their physiological functions and solve the problem of ranking conflicting motivations. The same mechanism operates in conflicts of biological and non-biological motivations, e.g., gustatory palatability versus money. The hedonic dimension of the conscious experience serves as the common currency allowing the ranking of priorities. Thus, maximization of pleasure solves the conflicts of motivations and optimizes behavior. It follows that the laws of sensory pleasure also apply to other hedonic experiences, but that is beyond the scope of the present chapter, which focuses on sensory pleasure and physiology.

Temperature

Skin sensation. Skin temperature produces a sensation. This obvious point is self-evident. We shall see that this dimension depends not only on the skin but also on inputs from the core temperature and the thermoregulatory set-point, i.e., where comfort and discomfort originate.

Influence of core temperature. The strong influence of core temperature on perception of thermal comfort is evidenced by reports from subjects whose core temperature has been modified while their mean skin temperature remains stable. Thus, when core temperature has deviated from normal, subjects complain of thermal discomfort although their skin temperature remains neutral. The primary component of such discomfort is the unpleasant sensation felt in the skin. The same cold temperature will feel unpleasant to a hypothermic subject and very pleasant to a hyperthermic subject. Thus, the pleasure or displeasure aroused by a thermal stimulus will depend on the subject's internal state. This phenomenon is therefore an alliesthesia similar to the example of sweet candies above (Figure 2.3). The internal signal that gives rise to an alliesthesic change is the difference between set-point temperature and actual core temperature. Two models have been proposed to predict preferred skin temperature (T_p):

$$T_p = a(T_c-b)T_{mean\ sk}+c \text{ (Cabanac, Massonnet, and Belaiche, 1972)}$$

$$T_p = a+bT_c+cT_{mean\ sk} \text{ (Bleichert, Behling, Scarperi, and Scarperi, 1973)}$$

and one predicting the subjective assessment (SA):

$$SA = aT_c + b dT_{c/dt} + cT_{mean\ sk} + e dT_{mean\ sk/dt} + fS + g \text{ (Attia and Engel, 1982)}$$

where a, b, c, e, f, and g are constants, T_c is core temperature, $T_{mean\ sk}$ is mean skin temperature, and S is a shivering factor (0 or 1). One model was proposed to predict alliesthesia:

$$a = f(T_c - T_{set})T_{mean\ sk}, \text{ (Attia and Engel, 1981)}$$

where T_{set} is the set-point temperature of the biological thermostat. Alliesthesia will affect different locations on the skin's surface simultaneously, a phenomenon that accounts for comfort and discomfort in various circumstances. For example, comfort persists when one side of the body is cooled and the other side warmed, or when the body receives asymmetrical radiation producing a difference of up to 13°C between front and back skin temperatures.

Set-point. The above signals and the proposed equations are also those that govern autonomic temperature regulation. This regulation possesses an additional implicit signal: The set-point. The actual internal temperature is compared to the set-point "wanted" by the organism. The activating signal for the regulatory responses, the "error signal," is the difference between the actual temperature and the set-point. When an error signal is detected, the organism produces the available corrective responses. We have seen that the pleasure of thermal sensation varies with hypo- or hyperthermia. Hyper- and hypothermia are defined as deviations of core body temperature from the set temperature, approximately 37°C in healthy subjects. In two cases, the set-point is cyclically reset.

Figure 2.2 Left – hedonic responses to temperature stimuli between pain thresholds (15 and 45°C) of a subject whose hand has been stimulated for 30 seconds, and whose body has been immersed in a well-stirred bath. Each dot is the response to a stimulus. Triangles = cold bath; circles = warm bath; hypothermic subject = open dots; hyperthermic subject = solid dots. It can be seen that core temperature determined the hedonic experience; e.g., to the hyperthermic subject, cold stimuli felt pleasant, and warm stimuli unpleasant. Right: Same legend as left except that the subject's core temperature was the same (ca. 38°C) for both solid and open dots. Solid dots indicate the responses when the subject was feverish, and open dots the responses from control session without fever. It can be seen that the modified set-point was sufficient to change the subject's thermal preference, all other conditions being identical. From Cabanac, 1969.

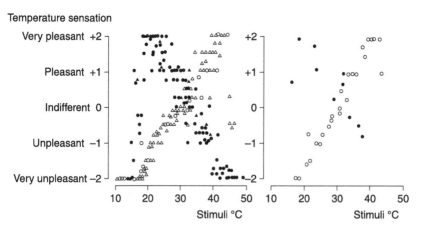

Temperature sensation

In the nycthemeral cycle, the set-point oscillates over a period of 24 hours. During this cycle, the pleasure of thermal sensation is precisely adjusted to maintain the oscillating set-point (Cabanac, Hildebrandt, Massonnet, and Strempel, 1976). Through this hedonic experience, behavior is in turn adapted to maintenance of the oscillating set-point; the preferred ambient temperature and dressing behavior will anticipate the oscillation. Similar oscillations of alliesthesia and dressing behavior occur during the ovarian cycle and maintain the 28-day hormonal resetting of the set-point (Cunningham and Cabanac, 1971).

Olfaction and Taste

Food stimuli. Pleasure aroused by eating shows an identical pattern. A given alimentary flavor is described as pleasant during hunger and becomes indifferent or unpleasant during satiation (Cabanac, Minaire, and Adair, 1968; Fantino, 1984, 1995) (Figure 2.3). Thus, eating is another example where the seeking of sensory pleasure leads to a behavior adapted to the physiological need of the body. Measurement of human feeding behavior confirms the above relationship of behavior with pleasure. It has been repeatedly demonstrated in the case of food intake (Fantino, 1984) that human subjects tend to consume foods that they report to be pleasant and to avoid those that they report to be unpleasant. Pleasure also shows a quantitative influence: The amount of pleasurable food eaten is a function of dietary restrictions, and increases after dieting.

Set-point. The similarity of pleasure from food and temperature stimuli was recognized when alliesthesia to sweet stimuli disappeared in

Figure 2.3 Pleasure (positive ratings) and displeasure (negative ratings) reported by a fasted subject, on an empty stomach, in response to the same taste stimulus, a sample of water containing sucrose presented repeatedly every third minute. Solid dots = the subject spat out the samples after tasting; open symbols = the subject swallowed the samples and thus accumulated a heavy sucrose load in his stomach. In the latter case, it can be seen that the same sweet taste that aroused pleasure in the subject when he had not eaten aroused displeasure when the subject was satiated. From Cabanac, 1971.

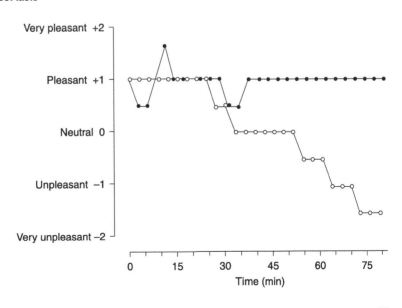

25

subjects whose body weight was reduced with dieting, and recovered when body weight returned to control value (Figure 2.4). When body weight was reduced, the error signal between actual body weight and set-point body weight inhibited the internal signal for alliesthesia. As a result, the ingestion of the same glucose load was not followed by a negative alliesthesia for alimentary stimuli. When negative alliesthesia is thus delayed, subjects tend to eat more as their satiety is retarded. The amount eaten is larger than when the subject is at set-point and the subject regains weight. When the body weight recovers its initial value and the subject is at set-point, alliesthesia and satiety return to normal and food intake tends to decrease. Thus food pleasure is related to regulation of the ponderostat

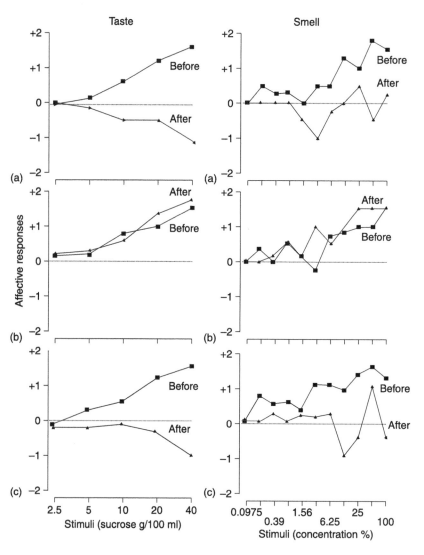

Figure 2.4 Verbal responses describing the hedonic experience of two subjects receiving sucrose stimuli (left) and orange odors (right). In all six boxes the subjects were tested before and after receiving a concentrated glucose load in their stomach, 50 g in 200 ml. The upper two boxes were control sessions on *ad-libitum* fed subjects. It can be seen that the glucose loads rendered the food stimuli, sweet taste, and orange odor unpleasant. In the middle boxes the subjects had lost weight (5.2 kg, left; 3.0 kg, right) through drastic low-caloric diets over several weeks. In lean subjects the gastric load no longer had any satiating effect. Several weeks later (lower boxes), the *ad-libitum* fed subjects returned spontaneously to their original weights; gastric load was effective again in producing an unpleasant sensation (negative alliesthesia). From Cabanac, Duclaux, and Spector, 1971.

in a way similar to that of thermal pleasure in regulation of the thermostat. Pleasure and displeasure from food stimuli not only are a driving force for adaptive behaviors, but also maintain the stability of the *milieu intérieur* as defined by the physiological set-points.

Conflicts Between Motivations

If sensory pleasure indicates physiological usefulness it would be of interest to explore situations with simultaneous and possibly conflicting motivations. McFarland and Sibly have pointed out that behavior is a *final common path* used in competition by all existing motivations. Only the most urgent at any time has access to behavior. To achieve the trade-off between motivations there is need of a *common currency* in order to rank the respective urgencies (McFarland and Sibly, 1975). Pleasure could be such a common currency. Indeed, pleasure is not limited to sensory pleasure. Several experiments were conducted to explore such a situation where one motivation was pitted against another, e.g., sweetness *vs.* sourness (Cabanac and Ferber, 1987), temperature discomfort *vs.* fatigue (Cabanac and LeBlanc, 1983), and chest fatigue *vs.* leg fatigue (Cabanac, 1986). In all of these experiments the subjects' behaviors were repeatedly consistent. In the bi-dimensional sensory situations imposed by the experimenters the subjects' perceptions and ratings were used to draw maps of bi-dimensional pleasure. When offered an unrestricted choice, the subjects tended to move to the areas of maximum pleasure in these maps. The subjects tended to maximize the algebraic sum of their sensory pleasure, or to minimize their displeasure. Thus, subjects in situations of conflicting motivations tend to maximize their sensory pleasure as perceived simultaneously in both dimensions explored. In addition, such an algebraic sum corresponds to the highest optimal point for the two opposing physiological functions in the conflict; e.g., in the conflict chest fatigue *vs.* leg fatigue, the minimization of displeasure produced a muscular work at constant power.

The Problem of Sensory Pleasure in Animals

Sensory pleasure is so hardwired that it irresistibly calls for an investigation into its phylogenetic origins. If maximization of sensory pleasure drives behaviors adapted to human physiological needs, given that animals also display such behaviors, one may also question whether animals experience sensory pleasure and behave in such a way as to maximize it. Anecdotally, when a pet or a farm animal seeks a sensory reward, whether food or petting, we spontaneously accept that the experience of pleasure drives that behavior. It is, nevertheless, necessary to try to obtain more objective evidence of such experience.

Temperature in Rats

Weiss and Laties (1961) have shown that in a cold environment rats will bar-press repeatedly to obtain bursts of infra-red heat. Such a behavior appears to replicate that of humans who seek heat when hypothermic. However, because peripheral heat warms the animal such a behavior might be pure reflexive temperature regulation without being necessarily triggered by a desire to maximize pleasure. Some light has been shed on this problem by Satinoff, when her rats bar-pressed for infra-red heat after their hypothalamus was cooled (Satinoff, 1964). In another study, rats were offered the possibility to either warm or cool their hypothalamus or their skin while either the hypothalamus or the skin was heated or cooled (Corbit, 1973; Dib, Cormarèche-Leydier, and Cabanac, 1982). However, a doubt still remained: Because cooling the skin or the central sensors triggers production of shivering, it cannot be ruled out that the thermoregulatory behavior displayed by the rats was a response to the shivering muscles rather than to an unpleasant skin sensation. The answer to this question was provided by a study on curarized hypothermic rats, which slowed or accelerated their heart rate in order to activate a heater without shivering (Cabanac and Serres, 1976). It is therefore highly likely that rats experience thermal pleasure and displeasure in the same way as humans do. Further, the responses to cooling and warming the hypothalamus and spinal cord would tend to show that the physiological bases of such pleasure are the same as those found in humans.

Taste in Rats

Ingestion of a gastric glucose load by human subjects makes sweet stimuli unpleasant if they are pleasant before the load (negative alliesthesia). Extrapolation of these results to animals has been made possible by the technique developed by Grill, Norgren, and Berridge (Berridge, 1996; Berridge and Grill, 1984; Berridge, Grill, and Norgren, 1981; Grill and Norgren, 1978) who have discovered that rats' behavior reflects the nature of the taste stimulus they receive in their mouths. When sucrose is injected into their mouths, rats show feeding responses, including movements of the upper lip, protrusion of the tongue, chewing movements, paw licking, etc. When quinine is injected into their mouths, rats present avoidance responses including gaping, dribbling, forward protruding of the head, etc. It must be remembered that the rats' feeding and avoidance responses are not food intake. They are mere reflexes that can occur in decerebrated rats (Grill and Norgren, 1978). However, since humans also exhibit the hedonic dimension of the taste sensation (Steiner, 1977), it may be hypothesized that the feeding/avoidance responses displayed by rats indicate of pleasure or displeasure as they do in humans. Actually, these facial and postural responses result from the same signals, including taste, ingestion, body weight set-point, etc., as those that trigger taste alliesthesia in human subjects (Cabanac and Lafrance, 1990). Therefore, taste pleasures not only exist in rats, but are adapted to the maintenance of the body's integrity and the operation of the ponderostat, i.e., the same pattern as that found in humans.

Another example of sensory pleasure applied to the defense of the *milieu intérieur* may be found in taste aversion, a mechanism that prevents poisoning. Garcia, Hankins, and Rusiniak, 1974 have discovered a third mechanism of acquisition, taste aversion, after Pavlovian reflex and operant conditioning. This mechanism is especially well suited to the defense of body integrity. Rats and other animals avoid the tastes of new baits that were followed with digestive disease. If the first encounter of the animal with a new flavor is followed over the next 24 hours with either nausea or diarrhea the animal will never consume it again. It is likely that this new flavor now arouses displeasure in the rat. Innumerable anecdotal evidence of humans who dislike a given flavor after having had nausea (such as in the first months of pregnancy) or diarrhea in the hours following a new taste, would tend to indicate that this is the case.

Conflicts Between Motivations in Rats

Is it possible to obtain empirical evidence, similar to that obtained in humans, showing that animals will seek sensory pleasure and succeed in trading off some amount of displeasure for it? In the obstruction method (Warden, 1931) in contrast to the operant conditioning method, the strength of a motivation is measured not as an operant or motor response, but rather as the decision by an animal to overcome an obstacle to obtain a reward.

Such a situation can be explored in the laboratory under conditions close to nature. Rats were trained for several weeks to feed each day from 10:00 am to 12:00 noon, i.e., two hours a day. Over the same period, they learned in a zigzag maze that, once a week, additional highly palatable food was available during the regular feeding session, 16 meters away from their shelter. On the day of an experimental session their shelter was kept warm but the maze and the bait were in a very cold environment of −15°C in turbulent air, a potentially lethal environment for a rat. Although regular laboratory chow was available *ad libitum* in their warm shelter, thus making foraging unnecessary, the rats invariably ran for short meals to the cold feeder to obtain the highly palatable food and rushed back to their warm shelter between the meals (Cabanac and Johnson, 1983) (Figure 2.5). With such foods, the rats

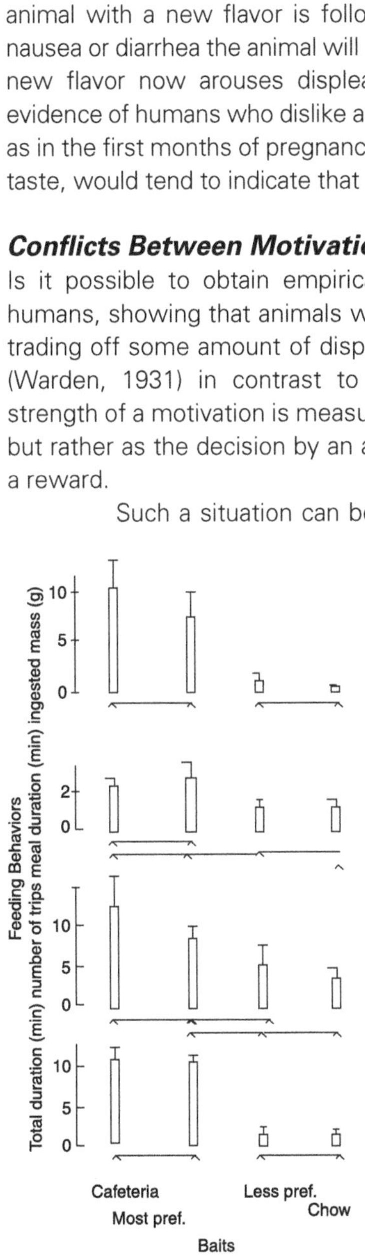

Figure 2.5 Rats make a tradeoff between cold discomfort for palatability. This figure shows the characteristics of rats' meals taken from a feeder (*Comportement au restaurant*) at −15°C placed 16 m away from the rats' warm shelter. The rats had water and chow *ad libitum* in their shelter and were offered either cafeteria diet, the most and the least favored bait of each animal, or lab-chow at the feeder (*Nourriture*). Bars above columns indicate S.E.; x___x lines join columns when not significantly different. Ingested mass is the amount ingested at the feeder in addition to the chow ingested in the warm shelter (not indicated on the figure); mean meal duration shows the time spent at the feeder in the severely cold environment. Total time at the feeder was obtained from direct recording. The columns on the left are higher than on the right; this shows that rats ventured repeatedly into the cold for highly palatable, but not for lowly palatable food. From Cabanac and Johnson, 1983.

consumed up to half their nutrient intake on that day in a potentially lethal environment (if they were to stay in it). For less palatable foods, the rats went only once or twice to the feeder over the two-hour session, and stayed for a shorter time. Although they did not become hypothermic, the rats gave signs that the environment was unpleasant: They rubbed their ears and areas of naked skin, and the tips of their ears and tails sometimes had been damaged by frostbite. Thus, the animals faced the unpleasant or even painful cold not out of necessity, since regular food was provided at no cost in their warm shelter, but for the pleasure of ingesting a palatable bait. This shows that they quantitatively matched the pleasure of food with the displeasure of enduring cold and, therefore, experienced pleasure.

Thus, we may accept that mammals experience sensory pleasure and that, like humans, their sensory pleasures are adapted to the fulfillment of physiological needs. Yet, humans and other mammals are not the only animals with behavior. On the contrary, as already stated, behavior in all animals serves physiology and further, the simpler the animal, the more dependent it is on its behavior. Therefore it is of interest to explore further this evolutionary sequence to detect signs of pleasure in animals more primitive than mammals.

Conflicts Between Motivations in Lizards

When confronted with an aversive and potentially lethal environment lizards modify their behavior so as to increase the number of their foraging expeditions while decreasing the duration of each one. This behavior allows them to maintain their core temperature and, at the same time, to consume a constant amount of food even when the environment's temperature has become dangerously cold (Cabanac, 1985). When iguanas are placed in the same situation – a cold environment with a warm shelter, but with regular food available in a warm corner of their terrarium – they forage for palatable lettuce less often with decreasing ambient temperature (Balaskó and Cabanac, 1998) (Figure 2.6). Therefore, reptiles will venture into a cold environment if necessary, but will not if highly palatable food is the sole motivation. These behaviors indicate that reptiles experience pleasure and that sensory pleasure is the signal for behaviors adapted to the maintenance of survival and body integrity.

Birds

If present-day mammals and reptiles experience sensory pleasure, birds too should possess this mental capacity because all three classes of vertebrates share common ancestry. Do birds experience sensory pleasure?

To answer this question I trained an African Grey parrot (*Psittacus erythacus*) named "Aristote" to talk. The first step consisted in gaining Aristote's affection. Then Aristote was taught to speak, following Irene

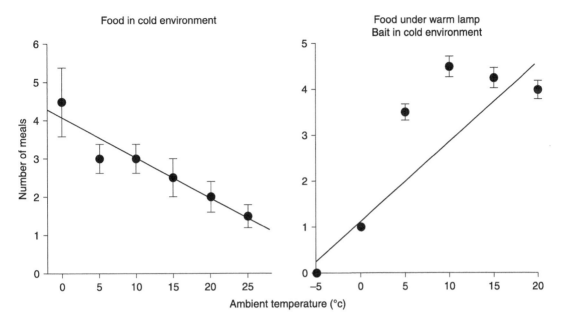

Figure 2.6 Feeding behavior of lizards. Left, four *Tupinambis teguixin* were placed in a terrarium with an infra-red lamp in a corner and food in the opposite corner of the terrarium. Thus they had to feed at a distance from the warm refuge. The mean number of visits to the bait by the lizards is plotted against the ambient temperature. When the temperature at the feeder was lowered the lizards modified their feeding pattern: They shortened the duration of their meals (not shown) and increased the number of their meals. As a result the feeding duration and the amount ingested remained steady (Cabanac, 1985). Right, three *Iguana iguana* lived in similar conditions but they had access to regular chow at their warm refuge and palatable lettuce was placed at a distance.

Pepperberg's triangular method: Another person and I would talk to one another and would look at Aristote only when it used understandable French words. Thus Aristote learned to say a few words to obtain toys or get my attention; e.g. "*donne bouchon*" (give cork) or "*donne gratte*" (give scratch), with the appropriate reward.

Lastly, the word "*bon*" (good) was added to the short list of words used by Aristote. I said "*bon*" whenever Aristote obtained the stimuli it had requested; e.g. "*gratte bon*". Aristote started to use short expressions such as "*yaourt bon*" (yogurt good). Finally, Aristote transferred the word "*bon*" to new stimuli such as "*raisin*" (grape), an expression I had never used myself. Such a transfer likely shows that this bird experienced sensory pleasure.

Long-term Regulation

Behavior is also adapted to the defense of homeostasis in the long term. In the experiments described above, the behaviors, food intake, temperature regulation, which were triggered by sensory pleasure, were not limited to correcting immediate needs but also anticipated future needs for chemical and thermal energy. Such anticipation is especially the role of the set-points. The set-points of the various regulations taking place in physiology are not fixed constants. On the contrary they tend to be permanently adjusted according to peripheral signals, or genetically programmed signals. The mechanism involved is a simple resetting with time. We saw above that the sensory pleasure aroused in temperature sensation, in humans, is adapted to the thermoregulatory set-point that oscillates over the nycthemeron (Cabanac et al., 1976). Numerous other examples can be found of such adjustments of various set-points. Among these, low palatability of food lowers the body weight set-point (Cabanac and Rabe, 1976), a mechanism that is especially well suited to the food environment available to our hunter-gatherer ancestors and anticipates irregular access to food. On the other hand palatable signals activate the physiological response before ingestion, a typical case of anticipation: Oral stimulation with sweet solutions causes a rapid elevation of blood glucose concentration in fasted rats; in addition their respiratory quotient simultaneously rises abruptly (Nicolaïdis, 1969). Emotion lowers the body weight set-point (Michel and Cabanac, 1999), thus saving behavior for fight or flight, and raises the temperature set-point (Briese and Cabanac, 1991), thus improving muscular work. During muscular exercise the ponderostat is reset at a lower value (Cabanac and Morrissette, 1992) in such a way that behavior may be devoted to its present task and not distracted with the motivation to eat. The ponderostat also follows the ovary endocrine female signals: In 350-gram female rats the threshold for hoarding food, i.e., the body weight set-point, is reset at a value higher after ovulation than before ovulation (Fantino and Brinnel, 1986). The fluctuation is ca. 10 percent of body weight; one may easily see how a raised set-point anticipates the future needs of pregnancy. Finally, in hibernating animals the ponderostat is set according to a sinusoid over the year (Canguilhem and Marx, 1973; Mrosovsky and Fisher, 1970), a mechanism that allows the storage of fat and the long wintering without food.

The long-term adjustments of the various set-points that adapt physiology and its behavioral response to changing needs have been termed by *homeorhesis* by Nicolaïdis, 1977 and *rheostasis* by Mrosovsky, 1990. Mrosovsky has superbly theorized on these long-term adjustments of the various set-points of the physiological regulations. He reached the concept of rheostasis by taking into account that (1) nocturnal hypothermia anticipates the next day's heat; (2) the longer-term fluctuations of the various physiological "constants" are indeed defended; (3) the lack of stability in the *milieu intérieur* is a specialized adaptation. This mechanism is close to the concept of *allostasis* created by Sterling and Eyer, 1981, in that it accounts for the long term. However, *rheostasis/homeorhesis* reinforces strongly the concept of set-point, contrary to *allostasis*, which, in my opinion, blurs the understanding of regulatory process.

Conclusion

Sensory pleasure is a very ancient phylogenetic mechanism, present in reptiles, birds, and mammals. Behavior is motivated by the trend to seek pleasure and to avoid displeasure. The hedonic dimension of sensation, therefore, is the motor of behavior. Experimental evidence shows that sensory pleasure takes place when peripheral stimuli tend to close the gap between actual internal states and their ideal set-points. Thus, maximization of sensory pleasure optimizes behavior, in terms of maintaining physiological integrity and anticipating future need. Such a mechanism allows one to select the optimal stimuli from a range of environmental conditions and the best compromise in the event that motivations enter into conflict with each other. Pleasure is an ancient mechanism that emerged with the early reptile ancestors to present-day reptiles, birds, and mammals. The remaining of pleasure by natural selection through evolution is another indirect indication of its beneficial function.

References

Attia, M. (1984). Thermal pleasantness and temperature regulation in man. *Neuroscience and Biobehavioral Reviews*, 8, 335–343.

Attia, M. and Engel, P. (1981). Thermal alliesthesial response in Man is independent of skin location stimulated. *Physiology & Behavior*, 27, 439–444.

Attia, M. and Engel, P. (1982). Thermal pleasantness sensation: an indicator of thermal stress. *European Journal of Applied Physiology*, 50, 55–70.

Balaskó, M. and Cabanac, M. (1998). Behavior of juvenile lizards (*Iguana iguana*) in a conflict between temperature regulation and palatable food. *Brain Behavior and Evolution*, 52, 257–262.

Bernard, C. (1855). *Leçons de physiologie expérimentale appliquée à la médecine. Cours du semestre d'été*, Vo. II. Paris: Baillière.

Berridge, K. C. (1996). Food reward: brain substrates of wanting and liking. *Neuroscience and Biobehavioral Reviews*, 20, 1–25.

Berridge, K. C. and Grill, H. J. (1984). Isohedonic tastes support a two-dimensional hypothesis of palatability. *Appetite*, 5, 221–231.

Berridge, K. C., Grill, H. J., and Norgren, R. (1981). Relation of consummatory responses and preabsorptive insulin release to palatability and learned taste aversions. *Journal of Comparative and Physiological Psychology*, 95, 363–382.

Bleichert, A., Behling, K., Scarperi, M., and Scarperi, S. (1973). Thermoregulatory behavior of man during rest and exercise. *Pflügers Archiv*, 338, 303–312.

Bligh, J. (1973). *Temperature regulation in mammals and other vertebrates*. Amsterdam: North Holland.

Briese, E. and Cabanac, M. (1991). Stress hyperthermia: Physiological arguments that it is a fever. *Physiology & Behavior*, 49, 1153–1157.

Cabanac, M. (1969). Plaisir ou déplaisir de la sensation thermique et homéothermie. *Physiology & Behavior*, 4, 359–364.

Cabanac, M. (1971). Physiological role of pleasure. *Science*, 173, 1103–1107.

Cabanac, M. (1985). Strategies adopted by juvenile lizards foraging in a cold environment. *Physiological Zoology*, 58, 262–271.

Cabanac, M. (1986). Performance and perception at various combinations of treadmill speed and slope. *Physiology & Behavior*, 38, 839–843.

Cabanac, M. (1992). Pleasure: the common currency. *Journal of Theoretical Biology*, 155, 173–200.

Cabanac, M. (1996). On the origin of consciousness, a postulate and its corollary. *Neuroscience and Biobehavioral Reviews*, 20, 33–40.

Cabanac, M. and Ferber, C. (1987). Pleasure and preference in a two-dimensional sensory space. *Appetite*, 8, 15–28.

Cabanac, M. and Johnson, K. G. (1983). Analysis of a conflict between palatability and cold exposure in rats. *Physiology & Behavior*, 31, 249–253.

Cabanac, M. and Lafrance, L. (1990). Postingestive alliesthesia: The rat tells the same story. *Physiology & Behavior*, 47, 539–543.

Cabanac, M. and LeBlanc, J. (1983). Physiological conflict in humans: fatigue *vs* cold discomfort. *American Journal of Physiology*, 244, R621–R628.

Cabanac, M. and Morrissette, J. (1992). Acute, but not chronic exercise lowers the body weight set-point in male rats. *Physiology & Behavior*, 52, 1173–1177.

Cabanac, M. and Rabe, E. F. (1976). Influence of a monotonous food on body weight regulation in humans. *Physiology & Behavior*, 17, 675–678.

Cabanac, M. and Serres, P. (1976). Peripheral heat as a reward for heart rate response in the curarized rat. *Journal of Comparative and Physiological Psychology*, 90, 435–441.

Cabanac, M., Duclaux, R., and Spector, N. H. (1971). Sensory feedbacks in regulation of body weight: is there a ponderostat? *Nature*, 229, 125–127.

Cabanac, M., Hildebrandt, G., Massonnet, B., and Strempel, H. (1976). A study of the nycthemeral cycle of behavioural temperature regulation in man. *Journal of Physiology*, 257, 275–291.

Cabanac, M., Massonnet, B., and Belaiche, R. (1972). Preferred hand temperature as a function of internal and mean skin temperatures. *Journal of Applied Physiology*, 33, 699–703.

Cabanac, M., Minaire, Y., and Adair, E. R. (1968). Influence of internal factors on the pleasantness of a gustative sweet sensation. *Communications on Behavioral Biology Part A*, 1, 77–82.

Canguilhem, B. and Marx, C. (1973). Regulation of the body weight of the European hamster during the annual cycle. *Pflügers Archiv*, 338, 169–175.

Cannon, W. B. (1932). *The wisdom of the body*. New York: Norton.

Corbit, J. D. (1973). Voluntary control of hypothalamic temperature. *Journal of Comparative and Physiological Psychology*, 83, 394–411.

Cunningham, D. J. and Cabanac, M. (1971). Evidence from behavioral thermoregulatory responses of a shift in set point temperature related to the menstrual cycle. *Journal de Physiologie*, 63, 236–238.

Darwin, C. (1872). *The expression of the emotions in man and the animals*. Chicago: University of Chicago Press.

Dib, B., Cormarèche-Leydier, M., and Cabanac, M. (1982). Behavioural self warming and cooling of spinal canal by rats. *Physiology & Behavior*, 28, 489–496.

Fantino, M. (1984). Role of sensory input in the control of food intake. *Journal of the Autonomic Nervous System*, 10, 326–347.

Fantino, M. (1995). Nutriments et alliesthésie. *Cahiers de Nutrition et Diététique*, 30, 14–18.

Fantino, M. and Brinnel, H. (1986). Body weight set-point changes during the ovarian cycle: experimental study of rats during hoarding behavior. *Physiology & Behavior*, 36, 991–996.

Garcia, J., Hankins, W. G., and Rusiniak, K. W. (1974). Behavioral regulation of the milieu interne in man and rat. *Science*, 184, 824–831.

Grill, H. J. and Norgren, R. (1978). Chronically decerebrate rats demonstrate satiation but not bait shyness. *Science*, 201, 267–269.

Hardy, J. D. (1971). Thermal comfort and health. *ASHRAE J.*, 77, 43–51.

Helmoltz, H. (1879). *Die Tatsachen in der Wahrnemung*. Berlin: Hirschwald.

Hess, W. R. (1936). Hypothalamus und die Zentren des autonomen Nervensystems: Physiologie. *Archiv für Psychiatrie und Nervenkranken*, 104, 548–557.

Krebs, J. R. and Davies, N. B. (1981). *An Introduction to Behavioural Ecology*. Sunderland, MA: Sinauer Associates.

Langlois, P. (1902). La régulation thermique des poïkilothermes. *Journal de Physiologie*, 2, 249–256.

McFarland, D. J. and Sibly, R. M. (1975). The behavioural final common path. *Philosophical Transactions of the Royal Society*, 270, 265–293.

Michel, C. and Cabanac, M. (1999). Opposite effects of gentle handling on body temperature and body weight in rats. *Physiology & Behavior*, 67, 617–622.

Mrosovsky, N. (1990). *Rheostasis, the physiology of change.* New York: Oxford University Press.

Mrosovsky, N. and Fisher, K. C. (1970). Sliding set-points for body weight in ground squirrels during the hibernation season. *Canadian Journal of Zoology*, 48, 241–247.

Müller, J. (1834–1840). *Handbuch der Physiologie des Menschen für Vorlesungen* (2 vol.). Coblenz: Hölscher.

Nicolaïdis, S. (1969). Early systemic responses to orogastric stimulation in the regulation of food and water balance: functional and electrophysiological data. *Annals of New York Academy of Sciences*, 157, 1176–1203.

Nicolaïdis, S. (1977). Physiologie du comportement alimentaire. In P. Meyer (Ed.), *Physiologie humaine* (pp. 908–922). Paris: Flammarion.

Pavlov, I. (1966). *Essential works of Pavlov (M. Kaplan).* Toronto: Bantam Books.

Pfaffmann, C. (1960). The pleasures of sensation. *Psychological Review*, 67, 253–268.

Richter, C. P. (1936). Increased salt appetite in adrenalectomized rats. *American Journal of Physiology*, 115, 155–161.

Satinoff, E. (1964). Behavioral thermoregulation in response to local cooling of the rat brain. *American Journal of Physiology*, 206, 1389–1394.

Steiner, J. E. (1977). Facial expressions of the neonate infant indicating the hedonics of food related chemical stimuli. In J. E. Weiffenbach (Ed.), *Taste and development* (pp. 173–189). Bethesda, MD: US Dept. Health Education Welfare.

Stellar, E. (1954). The physiology of motivation. *Psychological Review*, 61, 5–22.

Sterling, P. and Eyer, J. (1981). Allostasis: a new paradigm to explain arousal pathology. In S. Fisher and J. Reason (Eds.), *Handbook of Life Stress, Cognition, and Health.* New York: John Wiley & Sons.

Thorndike, E. L. (1898). Animal intelligence: an experimental study. *Psychological Monographs*, 2(4).

Tinbergen, N. (1951). *The study of instinct.* Oxford: Clarendon Press.

Warden, C. S. (1931). *Animal Motivation, Experimental Studies on the Albino Rat.* New York: Columbia University Press.

Watson, J. B. (1919). *Psychology from the Standpoint of a Behaviorist.* Philadelphia: Lippincott.

Weiss, B. and Laties, V. G. (1961). Behavioral thermoregulation. *Science*, 20, 1338–1344.

Wundt, W. (1874). *Grundzüge der physiologischen Psychologie.* Leipzig: Englemann.

Young, P. T. (1959). The role of affective processes in learning and motivation. *Psychological Review*, 66, 104–123.

Chapter 3

Consciousness, Well-being and the Senses

Derek Clements-Croome

The Nature of Consciousness

> We do not understand how the mind works – not nearly as well
> as we understand how the body works, and certainly not well
> enough to design utopia or to cure unhappiness.
>
> (Pinker 1997)

How do the neural processes occurring in our brains while we think relate
to our subjective sensations? Crick and Koch (1997) believe that this is a
central mystery of human life. The fundamental question that needs to be
understood is the relationship between the mind and the brain. We are
conscious or aware of events central to our attention or concentration at
any one time. Often there are peripheral events that feature in only a
fleeting way in our consciousness unless they manage to distract us. The
ability to focus the concentration or alertness for a particular event, such as
the work we are undertaking, is an important issue when discussing
productivity. For high productivity we need high and sustained levels of
concentration centred on the task being carried out. There are many short-
term, medium-term and long-term factors that can contribute towards
lowering productivity and these include low self-esteem, low morale, an
inefficient work organisation, poor social atmosphere or environmental
aspects such as excessive heat or noise. Factors that lower productivity,
by distracting our attention and diluting concentration, include lethargy,
headaches and physical ailments. These factors all feature in surveys
carried out on building sickness syndrome (Abdul-Wahab 2011). Crick and

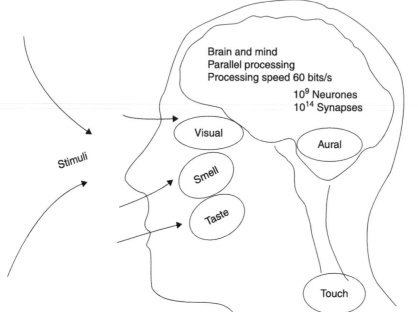

Figure 3.1 A representation of the human perceptual sensory system processing visual, sound, touch, smell and taste sensations.

Koch (1997) discuss visual consciousness in trying to reach an under-standing about how the brain interprets the visual world based on the information perceived by the visual system. Past experience evolved through living and from our genes features strongly in our responses to the environment around us. The stimuli from the environment trigger this system and arouse our consciousness to various levels of concentration.

The human perceptual sensory systems process information from visual, sound, touch, smell and taste sensations. Our surroundings create a sensory experience and hence affect the way we work. Conditions external to the body can disturb these systems; internal disturbances due to drugs or alcohol, for example, can also upset the response to the world around us. Greenfield (1997) believes that consciousness is impossible to define. She goes on to state that neuronal connectivity is a very important feature of the brain, which means that it is the connections rather than the neurons themselves that are established as a result of experience. It is the pattern of these experience-related connections that largely distinguishes the individuality of a human being. The pattern of these connections is continually evolving. It appears that the brain remains adaptable and sensi-tive to experience throughout a person's life. Consciousness changes as biorhythms and flows of hormones alter through daily cycles. Greenfield (1997) concludes that a critical factor could be the number of neurons that are gathered up at any one time, and it is this that determines one's consciousness.

The state of knowledge about environmental factors is uneven. It is probably fair to say that there is a high level of knowledge about how heat, light and sound affect our thermal, visual or auditory responses. There is much less information about how we react to combinations of these stimuli and also about how electromagnetic, geomagnetic and chemical fields affect the sensory system. An added complication is that human responses are partly physiological and partly psychological. This makes the measurement of responses difficult because objective measures of lighting or temperature levels are comparatively easy, but assessing people's judgements about preferred or acceptable levels of light and heat is much more difficult. Yet another complication is that reception of information from visual images, music or speech, smells or touch interact with one another. The sense organs extend beyond the eyes, ears, nose, mouth and skin and include the vestibular organs concerned with orientation, posture and locomotion, as well as a variety of respiratory and thermo receptors that respond to air quality, pressure and temperature.

Our response to the world around us occurs at various levels. For example, a cartoon can depict recognisable people from the skimpiest of outlines. The outline form and a few added clues about detail are all that is required to recognise the person being represented. Contrast this with a portrait by Rembrandt, for example, in which colour, texture and shading give much more detailed information that triggers higher orders of aesthetic and emotional response. Likewise, the quality of the environment has a basic structure upon which is superimposed more detail. For example, air movement can be represented at a basic level by a mean velocity, but a more complete picture would refer to the degree of turbulence, the peak as well as the mean velocity, and the periodicity of the air-flow. Crick and Koch (1997) describe various levels of representations that occur in the visual field. They suggest that there may be a very transient form or fleeting awareness that represents simple features and does not require an attention-awareness mechanism. The renowned psychologist William James believed that consciousness was not a *thing* but was *processed thought*, which involved attention and short-term memory. From brief awareness the brain constructs a view-centred representation and the visual inputs awaken a greater level of attention. Crick and Koch go on to suggest that this in turn probably leads to a three-dimensional object representation and thence to more cognitive ones.

In the design of the productive workplace an attempt is being made to set conditions that allow selected information to be perceived and transmitted quickly through the human perceptual sensory system. This pathway must not be trammelled by extraneous information from peripheral stimuli. Efficient and effective work processes and organisation, besides controllable environmental conditions, can help this process given that the person is healthy in mind and body and there is no interference at a social level from any other person.

Chalmers (1996) asks the following about consciousness:

- How can a human subject discriminate sensory stimuli and react to them appropriately?
- How does the brain integrate information from many different sources and use this information to control behaviour?
- How is it that subjects can verbalise their internal states?
- How do physical processes in the brain give rise to subjective experience?

Crick and Koch (1997) suggest that *meaning* is derived from the linkages among the various representations of the neuron firing fields that are spread through the cortical system in a vast network, equivalent to a huge database that is changing as the experiences of the individual increase throughout life. Changes bring about the process of learning. However, many questions remain unanswered. The existence of consciousness does not seem to be derivable from physical laws, and because consciousness is strongly subjective, there is no direct way to monitor it, although questionnaires and semi-structured interviews are techniques that are employed. Chalmers (1996) goes on to say that it is valid to use people's descriptions of their own experiences. There have been several surveys of productivity using self-assessment techniques, which is a scientifically acceptable procedure as long as the subject attempts to structure the information output in an objective fashion.

Allwright (1998) describes nine levels of consciousness and these define the various levels of sensory experience:

- the *five senses* felt by the eyes, ears, nose, mouth and skin;
- the *integration of senses* using reason and logic;
- *rational thought* expressed via self awareness and intuition;
- the *stores of experience* in the short- and long-term memories;
- *pure consciousness* within the inner self; this also involves emotion.

Although the five basic senses are often studied as individual systems covering visual, auditory, taste, smell, orientation and the haptic sensations, there is an interplay between the senses. For example, eyes want to collaborate with the other senses. All the senses can be regarded as extensions of the sense of touch, because the senses as a whole define the interface between the skin and the world. Various parts of the human body are particularly sensitive to touch. The hands are not normally clothed and act as our touch sensors. But the skin of the body reads the texture, weight, density and temperature of our surroundings. The combination of sight and touch allows the person to get a scale of space, distance or solidity. Allende (1998) refers to the sense of smell as so important in evoking memories. Tillotson (2004) describes the power of scent in arousing emotional as well as cognitive responses (see Figure 3.2). The Kajima building in Tokyo uses various aroma fragrances to condition the air and research shows these can help

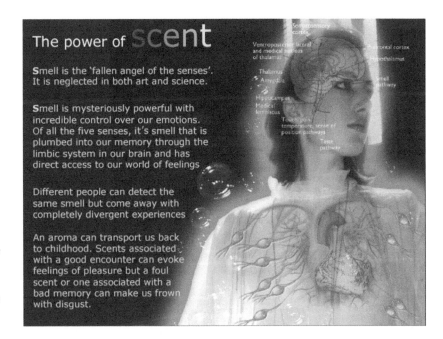

The power of scent

Figure 3.2 The complexity of sense and perception. Often what we perceive as a sense, like smell, is actually a perception that we link to the sense (Tillotson 2004).

people not just to feel a sense of freshness but also help them concentrate as well as offset fatigue (see Figures 3.3 and 3.4; Takenoya 2006).

Multi-Sensory Experience

Buildings should provide a multi-sensory experience for people and uplift their spirits. A walk through a forest is invigorating and healing due to the interaction of all sense modalities; this has been referred to as the

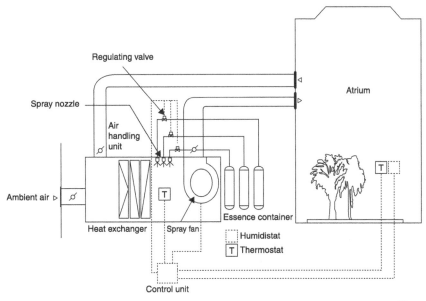

Figure 3.3 Atrium fragrance air-conditioning system in Kajima building in Tokyo. Source: Takenoya in Clements-Croome 2006.

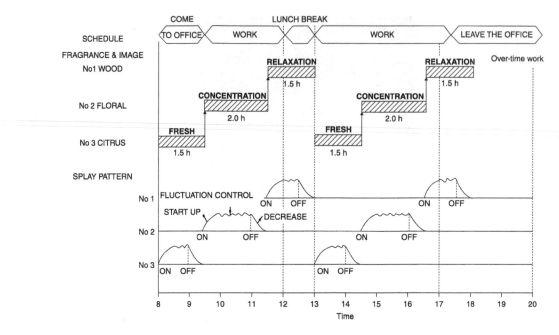

Figure 3.4 Atrium Fragrance and Control Scenario for Kajima Building in Tokyo. Source: Takenoya in Clements-Croome 2006.

polyphony of the senses. One's sense of reality is strengthened and articulated by the interaction of the senses. Architecture is an extension of nature into the manmade realm and provides the ground for perception, and the horizon to experience from which one can learn to understand the world. Buildings filter the passage of light, air and sound between the inside and outdoor environments; they also mark out the passage of time by the views and shadows they offer to the occupants. Pallasmaa (1996) gives an example to illustrate this point. He believes that the Council Chamber in Alvar Aalto's Säynätsalo Town Hall recreates a mystical and mythological sense of community where darkness strengthens the power of the spoken word. This demonstrates the very subtle interplay between the senses and how environmental design can heighten the expression of human needs within a particular context.

The surfaces of the building set the boundaries for sound. The shape of the interior spaces and the texture of surfaces determine the pattern of sound rays throughout the space. Every building has its characteristic sound of intimacy or monumentality, invitation or rejection, hospitality or hostility. A space is conceived and appreciated through its echo as much as through its visual shape, but the acoustic concept usually remains an unconscious background experience. Architecture emancipates us from the embrace of the present and allows us to experience the slow healing flow of time. Again, buildings provide the contrast between the passing of history and the time scales of life.

Proust gives a poetic description of a space of intimate warmth next to a fireplace sensed by the skin: '*it is like an immaterial alcove, a warm cave carved into the room itself, a zone of hot weather with floating boundaries'*. There is a subtle transference between tactile, taste and temperature experiences. Vision can be transferred to taste or temperature senses; certain odours, for example, may evoke oral or temperature sensations. The remarkable world-famous percussionist Evelyn Glennie is deaf but senses sound through her hands and feet and other parts of her body. Marble evokes a cool and fresh sensation. Architectural experience brings the world into a most intimate contact with the body. The body knows and remembers. The essential knowledge and skill of the ancient hunter, fisherman and farmer, for example, can be learnt at a particular time but, more importantly, the embodied traditions of these trades have been stored in the muscular and tactile senses. Architecture has to respond to ways of behaviour that have been passed down by the genes. Sensations of comfort, protection and home are rooted in the primordial experiences of countless generations. The word 'habit' is too casual and passes over the sequels of history embedded within us. Isabel Allende describes the idea for her book *Aphrodite* (1998) as being *'a mapless journey through the regions of sensual memory'*. From early times the fire has been a symbol of human multi-sensory experience. The fire gave light in darkness; it produced warmth for the body and heat for cooking; it was a protection from hungry animals of prey; it was a social focus. A campfire today, as throughout history, enhances our well-being and uplifts the spirits.

Well-being

Well-being reflects one's feelings about oneself in relation to one's world. Maslow (1943) defined our basic physiological, psychological and social needs and if these are satisfied then the well-being of an individual is likely to be high. More recent work on the science of well-being is described by Huppert *et al.* (2005). The question of the well-being of the employee is considered by Warr (1998). Defining well-being is not easy because there are job-specific issues and also people have feelings and concerns that are not related to a particular environment. Warr (1998) proposes a view of well-being that comprises three scales: pleasure to displeasure; comfort to anxiety; enthusiasm to depression. There are job and outside-work attributes that characterise one's state of well-being at any point in time and these can overlap with one another. Well-being is only one aspect of mental health; other factors include personal feelings about one's competence, aspirations and degree of personal control. Ten features of jobs are described by Warr (1998) that have been found to be associated with well-being. He believes that stable personality characteristics as well as age and gender are also significant. Environmental determinants of well-being are described as: the opportunity for personal control; the opportunity for using one's skills; externally generated goals; variety; the environment; availability of money; physical security; supportive supervision; the opportunity for interpersonal contact and job status in society. Warr (1998) reviews work that indicates that greater well-being is significantly associated with better job performance, lower absenteeism and reduced probability of leaving employees. The organisation as well as personal factors are also important.

Heerwagen (1998) draws attention to work in organisational psychology that shows the relationship between buildings and worker performance (P) is interrelated as shown:

$$P = \text{Motivation} \times \text{Ability} \times \text{Opportunity}$$

An individual has to want to do the task and then has to be capable of doing it; last but not least resources and amenities have to be available so that the task can be done. The built environment provides physical and social ambience that affect motivation; the provision of individual control and a healthy environment can enable ability to flourish; communications systems, restaurants and other amenities aid motivation and ability even further by providing opportunity for task implementation.

We all have circadian rhythms, physiologically and psychologically, and these change as we carry out different activities during the course of a day. There is a large variation in these needs, and also behaviour patterns, between one individual and another. It is important that when people are within buildings they have contact with the outside world throughout their working day, and also have the means to adjust their environment according to changing needs. The built environment therefore has to be sensitive to these requirements and allow individuals to control their surroundings, as well as provide adaptability to changing needs.

Well-being and Productivity

Myers and Diener (1997) have been carrying out systematic studies about awareness and satisfaction with life among populations. Psychologists often refer to this as subjective well-being. Findings from these studies indicate broadly that those that report well-being have happy social and family relationships; are less self-focused; less hostile and abusive; and less susceptible to disease. It appears also that happy people typically feel a satisfactory degree of personal control over their lives, whether in the workplace or at home. It is probably fair to assume that it is more likely that the work output of a person will be high if their well-being is high. Jamison (1997) reviews research that links manic-depressive illness and creativity. Many artists such as the poets Blake, Byron and Tennyson, the painter Van Gogh and the composer Robert Schumann are well-known examples of manic depressives. The work output of such people is distinguished but lacks continuity. Mozart and Schubert were not classified as manic depressives and their work output was consistently high throughout their short lifespans. In contrast, Robert Schumann was very prolific during 1839–41 and 1845–53, whilst suffering hypomania throughout 1840 and 1849.

Between these time-spans he had suffered from severe depression and before 1838 and after 1853 made suicide attempts. In the workplace one is not expecting creativity at this level of genius; rather, it is hoped that there will be a consistently high standard of work performance.

It is interesting to consider some case studies of the competitors at the Mind Sports Olympiad held in London during August 1997 (Henderson 1997). One competitor could memorise a pack of 52 cards in just over 30 seconds. His daily routine involves running four miles a day; no alcohol for six weeks before a tournament, during which he eats a lot of pasta and other high-carbohydrate food to keep the blood sugar level high; he also takes regular doses of a Chinese herb called *Ginkgo biloba* to improve blood circulation; he practises the trance-like state that is needed to perform his memory feats and has regular brain scans. This competitor believes in lowering his brain activity to the optimum concentration level for this type of feat. This means that his brain activity rate is reduced to 5–7 Hz, which allows a higher degree of meditative-type concentration than the normal brain activity of 12–14 Hz. Another competitor stated that he had a meat-free diet and a fitness programme, ran marathons and played tennis matches and dived to improve concentration. Diving is about poise and balance, and requires the same sort of mental rigour that is needed for competitions or solo performances. Tony Buzan is one of the organisers of the Olympiad, and believes that mind-training techniques can open up a new sphere of mental fitness, which needs to be integrated with a physical fitness programme. The brain uses 40 per cent of the body's oxygen, and a healthy body promotes brain activity. Buzan goes on to say that the imagination can do for the mind what weight training can do for the body. Everyone can do concentration exercises almost anywhere by, for example, watching a vase of flowers and concentrating on every detail, then closing his or her eyes and imagining it.

People do not have to be Olympiad competitors to get more out of their work. Townsend (1997) states that 25 per cent of us enjoy our work but the rest of us do not. Productivity suffers as a consequence, due to the workplace being more a place of conflict and dissatisfaction. Lack of productivity shows up in many ways, such as absenteeism, arriving late and leaving early, over-long lunch breaks, careless mistakes, overwork, boredom, frustration with the management and the environment. In the same way as the Olympiad competitors aim to focus their mind completely on the task in hand, and believe we can all try to do this, and when we succeed the whole body feels different. Townsend (1997) goes on to say that people in the workplace can be encouraged to use both sides of their brain. The left-side part is concerned with logic, whereas the right-hand side is concerned with feeling, intuition and imagination (Ornstein 1973). If logic and imagination work together, problem solving becomes more enjoyable and more creative. Of course some people thrive on change while others prefer to do repetitive types of work. There seems to be no doubt that the industrial and commercial worlds can play a leading role in increasing the awareness of their workforce about these possibilities. It is also important to start educating and training our school children about these issues. Concentrated meaningful thinking needs practice and a deeper consideration than we normally give it.

References

Abdul-Wahab, S. A. (2011) *Sick Building Syndrome*. Berlin: Springer.

Allende, I. (1998) *Aphrodite: A Memoir of the Senses*. London: Flamingo.

Allwright, P. (1998) *Basics of Buddhism*. London: Taplow Press.

Chalmers, D. J. (1996) *The Conscious Mind: In Search of a Fundamental Theory*. Oxford: Oxford University Press.

Clements-Croome, D. J. (ed.) (2006) *Creating the Productive Workplace*, second edition. Abingdon: Spon-Routledge.

Crick, F. and Koch, C. (1997) The problem of consciousness, *Scientific American*, Special Issue, 'Mysteries of the Mind', January, 19–26.

Greenfield, S. (1997) How might the brain generate consciousness? *Communication Cognition*, 30, 3–4, 285–300.

Heerwagen, J. H. (1998) 'Productivity and Well-Being: What are the Links?' American Institute of Architects Conference on Highly Effective Facilities, Cincinnati, 12–14 March.

Henderson, M. (1997) Mental athletes tone their bodies to keep their minds in shape, *The Times*, 19 August, 6.

Huppert, F. A., Baylis, N. and Keverne, B. (2005) *The Science of Well-Being*. Oxford University Press.

Jamison, K. R. (1997) Manic Depressive Illness and Creativity, *Scientific American*, Special Issue, 'Mysteries of the Mind', January, 44–49.

Maslow, A. H. (1943) A Theory of Motivation, *Psychology Review*, 50, 4, 370–396.

Myers, D. G. and Diener, E. (1997) The Pursuit of Happiness, *Scientific American*, Special Issue, 'Mysteries of the Mind', January, 40–49.

Ornstein, R. E. (1973) *The Nature of Consciousness*. London: Viking.

Pallasmaa, J. (1996) *The Eyes of the Skin: Architecture and the Senses*. London: Academy Editions.

Pinker, S. (1997) *How the Mind Works*. London: Allen Lane.

Takenoya, H. (2006) Airconditioning Systems of the KI Building, Tokyo in D. J. Clements-Croome (ed.) *Creating the Productive Workplace*, second edition. Abingdon: Spon-Routledge.

Tillotson, J. (2004) 'Interactive "Scentsory Design" for Health and Well-being', The Institute of Nanotechnology conference on New Technologies and Smart Textiles for Industry and Fashion, London, 1–2 December.

Townsend, J. (1997) How to draw out all the talents, *The Independent*, 24 July, 17.

Warr, P. (1998) 'What is our Current Understanding of the Relationships between Well-Being and Work?' Economics and Social Sciences Research Council Seminar Series at Department of Organisational Psychology, Birkbeck College, London (ed. R. Briner), 22 September, and *Journal of Occupational Psychology* (1990) 63, 193–210.

Chapter 4

Cultural Responses
to Primitive Needs

Nick V. Baker

Introduction

Although we spend 95 per cent of our time indoors, we are really outdoor animals. The forces that have selected the genes of contemporary man are found outdoors in the plains, forests and mountains, not in air-conditioned bedrooms and at ergonomically designed workstations. Fifteen generations ago, a period of little consequence in evolutionary terms, most of our ancestors would spend the majority of their waking hours outdoors, and buildings would primarily provide only shelter and security during the hours of darkness. Even when inside, the relatively poor performance of the building meant that the indoor conditions closely tracked the outdoor environment.

Furthermore, many of the activities that played a vital role in survival demanded an intimate knowledge of the climate, the weather and the landscape. Agriculture is an obvious example; rainfall, frosts, wind and their interaction with the landscape – shelter, drainage, pests, etc. – constantly reinforced man's link with nature.

Robert Winston[1] points out that whilst it is commonly accepted that our physical attributes derive from our primitive ancestors, it is less widely recognised that our behavioural traits may also. He refers to these as instincts, and uses the primitive model to explain our emotions in relation to families, religion and society. By implication, he is saying that certain behavioural responses are genetically determined and that we could expect these responses to change only by evolutionary mechanisms with a timescale of many hundreds of generations. In this paper we extend the argument to explain our response to our present day built environment – which differs even more from our primitive origins than do our contemporary social and family structures.

It is an appealing thought that there is some deep and causal relationship between our adjustment of a thermostat and the action to take refuge from the winter night in a cave; between the tolerance of lower colour temperature light sources at night and the lighting of the multi-purpose fire at the entrance to the cave following sunset; between the multi-billion-pound industry in cut flowers and houseplants, and the daily surveillance by primitive man of natural landscape and vegetation. But is there any evidence to support this link, and if it is proven, how will this knowledge help us to understand and improve our modern environment?

Thermal Comfort

Our modern indoor lifestyle is consuming massive amounts of fossil energy, simply to isolate ourselves from the forces that moulded us. In the last 25 years we have consumed as much fossil energy as in the history of man. Undoubtedly our drive to engineer the environment is broadly the same urge that made the primitive hunter-gatherer first improve his cave, stockpile food and fuel, then cultivate plants, domesticate animals, form cooperative groups, trade and so on. It seems, though, that our techno-logical momentum has caused us to over-shoot; to deliver too much of a good thing, to interpret the life-saving instinct to mitigate the cold by throwing another log on the fire as the need to eliminate all thermal sensation at any cost.

To explore this question, we will first turn to the topic of thermal comfort, since both the physiological and behavioural aspects have been well researched. Temperature, or rather the heat balance of the body that it controls, is one of the key environmental parameters affecting survival. We would expect it to be one of the most vital responses hardwired in our genes. With civilisation and development it has lost nothing in its impor-tance, for in struggling to isolate ourselves from the natural variations in temperature, energy for heating and cooling buildings has become the largest single energy end use.

The most essential characteristic of the outdoor environment is its variability. There is variability on different time scales – daily and annual cycles as well as the quasi-random nature of weather – and on different scales of space ranging from human scale to global scale. The consensus view, that supports a massive heating, ventilating and air-conditioning industry, is that the engineer's proper mission is to provide a stable, opti-mised environment, independent of the natural world outside.

However, this is being challenged. More than 30 years ago, Lisa Heschong,[2] in her highly original work *Thermal Delight*, decried thermal uniformity. A decade later, this time as a result of rigorous field studies, Schiller[3] concluded that "people voting with extreme [thermal] sensations are not necessarily dissatisfied". Since then many field studies have confirmed that thermal variation is tolerated, and in many cases enjoyed.

Thermal Comfort – the Two Models

We have, then, two schools of thought – the conventional view that thermal comfort is best described by thermal neutrality brought about by a steady state heat balance, and those that believe that thermal comfort can be achieved within a range of thermal sensations, provided adaptive behaviour is possible. The former school, based on responses of subjects in climate chambers, is epitomised by the work of Fanger,[4] whilst the latter uses evidence from subjects in real buildings, typified by the work of Humphreys and Nicol.[5]

Few would suggest that either represents bad science, yet they seem to reach significantly different conclusions. Why is this? One explanation would come as no surprise, that people behave differently in different contexts. It is not surprising that the subject, who has been told how to dress, been told how to sit, been told what task to do, in an unfamiliar climate chamber with no windows to open or warm radiator to draw closer to, responds differently from a person working in their study at home. The latter knows that there is a cold beer in the fridge or a warm sweater if needed. This has been tested directly by Oseland,[6] who observed that the same group of subjects when tested in three contexts – climate chamber, workplace and their home – became progressively more tolerant, accepting winter comfort temperatures 3°K lower than in the climate chamber, with an intermediate value in the workplace.

The key difference between the climate chamber and the real working or living environment is that in the second case the subject has a range of actions available to him or her that will mitigate the non-neutral thermal sensation. We refer to these actions as *adaptive behaviour*, and the facility to carry them out as *adaptive opportunity*.

Field Studies

The role of adaptive behaviour in achieving thermal comfort has received considerable attention in the last few years and the importance of *adaptive opportunity* (Figure 4.1) has been identified by Heerwagen,[7] Bordass[8] and Baker and Standeven.[9] This is the real and perceived freedom to make adjustments to the local environment (open windows, deploy shades) or to one's own status (remove clothing, move to cooler part of the room, alter posture). Work by Guedes[10] shows that in a large sample of office workers in Portugal, occupants felt more satisfied with the thermal conditions where there were openable windows, *even when the opportunity for opening them was not taken.* This strongly suggests that there is a psychological as well as physical aspect to adaptive behaviour. We will return to this issue later.

Even in more extreme climates, adaptive actions are often sufficient to achieve thermal satisfaction for wide ranges of thermal conditions. In a study in the Sudan, Merghani[11] observed that occupants of courtyard houses utilised the spatial and temporal range of temperatures available in

(a)

(b)

Figures 4.1a and 4.1b Two office environments in Lisbon showing good (above) and poor (below) adaptive opportunity.

the rooms and courtyard to maximise their comfort. Figure 4.2 shows that the occupant is highly selective, the temperature chosen by an occupant remaining close to the predicted comfort temperature throughout the day.

The migration was not solely comfort seeking, but was also for practical and social functions. However, habitual use of the house in this

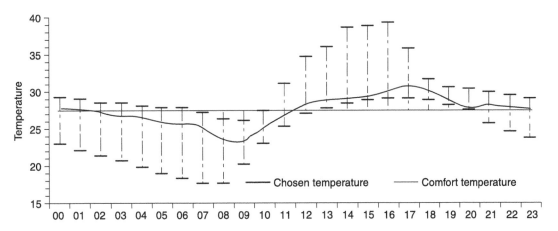

Figure 4.2 Chosen temperature (heavy line), i.e. the temperature at the location of the occupant throughout the day, with the range of temperatures existing in the building at each hour indicated by the bar. Note the chosen temperature follows the comfort temperature almost as closely as possible (see Merghani in Note 11 of this chapter).

way has a mutually reinforcing effect on these functions, and leads to increased overall satisfaction. This supports the notion that thermal comfort is a strongly contextual and holistic phenomenon. It also explains why under the closely controlled conditions of a test chamber, subjects respond in a completely different way.

Outdoor Comfort

In a study in Cambridge by Nikolopoulou,[12] it was found that people sitting outdoors in public places had greatly increased tolerance of non-neutral conditions, compared to what we would expect for indoor comfort. Typically, satisfaction was around 85 per cent compared with a predicted value of 35 per cent. Note that the predicted satisfaction, using Fanger's heat balance model, had already taken account of clothing level and metabolic rate, suggesting that there must be a strong psychological factor to account for the wide difference between the *predicted* and the *actual* satisfaction. She also observed significantly higher satisfaction when people were free to suit themselves when to leave, than when they were waiting to meet someone. This indicates that the element of choice has a significant and measurable effect.

The increase in tolerance was noticeably greater than is found inside buildings, even when they are regarded as having good adaptive opportunity. Could this be because the subjects are in outdoor and "natural" surroundings? Nikolopoulou also found that for subjects suffering from overheating discomfort in a sunlit street, where there was no natural landscape and little opportunity to seek shade, their increased tolerance was reduced.

These three examples certainly demonstrate adaptive behaviour, but they do not prove that it is essential for the environmental variance to which the subject responds to be "natural", although the increased tolerance in the latter case certainly points in that direction.

51

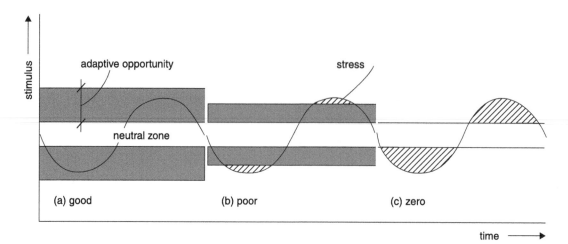

Figure 4.3 A general theory of environmental comfort. The neutral zone is effectively extended by the presence of adaptive opportunity. Note that in climate chamber experiments there is (by definition) no adaptive opportunity, which results in any departure from neutrality resulting in stress.

Evidence from thermal comfort field studies such as these, and studies in visual comfort (described below), has prompted us to propose a "general theory of environmental comfort". It relates the environmental stimulus (hot and cold in the case above), the degree of adaptive opportunity, and the resulting stress, as illustrated in Figure 4.3. It reconciles the discrepancy between the two approaches to thermal comfort and underlines the importance of providing adaptive opportunity rather than close environmental control.

The Luminous Environment

Our sensitivity to light is very different from our sensitivity to heat. Light in itself is rarely life threatening. However, in its role as a carrier of information, it may well become critical to survival. It is not difficult to think of cases where this is so, for both primitive and modern man. Natural light also signals the diurnal cycle of rest and activity, preparing the human for tasks that are most definitely critical to survival. It is well known that exposure to a natural cycle of daylight is instrumental in synchronising the body by the suppression of the hormones melatonin and seratonin.

But do we find responses to the luminous environment that are directly analogous to the thermal environment? A study carried out by Parpairi[13] in Cambridge showed an unexpected result. She studied user responses to different daylight conditions in a number of university libraries. Two cases are shown in Figures 4.4a and 4.4b: one, in a study carrel where the illumination is of high technical quality (glare-free diffuse light without the distraction of high contrast), and another close to the window where conditions varied strongly with the weather conditions and in particular the presence of sunlight.

Her findings show that the preferred condition was the second. Users found that they enjoyed the sunlit view of the River Cam, and if the

(a)

**Figures 4.4a and
4.4b** Comparison of
occupant response to
different daylight
environments in two
libraries.

(b)

glare became unbearable, they could retreat into a shaded part of the room
(seen on the left of the picture). The building offered adaptive opportunity
and in spite of strong stimuli of a natural origin, occupants reported a high
level of satisfaction. This case does seem to be closely analogous with
enhanced levels of thermal satisfaction under similar natural stimuli.

However, it is more complex because of the far greater information carrying capacity of light. It is interesting to speculate if the result would have been the same had the idyllic river scene been replaced by a car park or rubbish dump.

Clearly the information carried is important, even when it does not relate to the central task. We are dealing with *ambience* here, and it seems that ambience associated with nature is highly valued.

A striking and much quoted study carried out by Ulrich[14] investigated the impact of daylit views on patients recovering from surgery. He showed that patients recovered more rapidly when able to view a middle-distance natural scene including trees, than when viewing a blank wall (Table 4.1).

Even without a view, the dynamic quality of daylight seems to have an intrinsic value in the healing process. Keep, James and Inman[15] report on a comparison between the intensive care units at Plymouth and Norwich. It was found that patients from the Norwich unit, which was windowless, had a much less accurate memory of their length of stay, were subject to greater problems of disorientation, and recovered more slowly, although the windows at Plymouth were only translucent providing daylight, but no external view.

The intrinsic value of daylight in schools has been recognised for more than a century. In *School Architecture*, a manual prepared to assist in the design of urban schools following the passing of the Education Act in 1870, E.R. Robson[16] says rather poetically – "*It is well known that the rays of the sun have a beneficial influence on the air in the room . . . and are to a young child very what they are to flowers*".

Much more recently the value of daylight has also been quantified in the learning environment. In a study carried out by the Heschong Mahone Group[17] in the US, using data from government learning performance tests, it was shown that for eight- to ten-year-old children, annual progress in maths and English was improved from 6 to 26 per cent for daylit spaces. The effect was observed where daylight entered via diffuse

Table 4.1 Comparison of requested analgesic doses per patient for wall-view and tree-view patients; 46 patients between two and five days after surgery – see R.S. Ulrich in Note 14 of this chapter.

Analgesic strength	number of doses	
	wall group	tree group
Strong	2.48	0.96
Moderate	3.65	1.74
Weak	2.57	5.39

rooflights, but the largest progress was found where daylight also entered via windows.

Biophilia

"I love nature" is a phrase that covers a wide range of emotions. Responses may include a wide range of actions, from hill-walking holidays to subscriptions to the WWF, from designing buildings that look like insects, to keeping indoor plants in the office. But there is one thing in common with these diverse actions: all look outside the species of homo sapiens and its immediate self-constructed world, for some kind of inspiration or, at other times, solace. Is this just to be dismissed as sentimentality? Is nature to be seen important only if it yields raw materials for drugs, food and manufacture; is biodiversity to be protected simply for the utility of the gene pool, or is this engagement of deeper significance?

In developed nations, many of us live lives highly divorced from what we choose to consider as nature. However, almost wherever we look in the man-made world we still see references to nature somewhere. The question that this paper poses, then, is should we actively promote and provide this – nature reserves, parks, gardens, natural materials, climatically responsive buildings, down to indoor plants and pictures of distant mountains on the wall? Or should we simply respond to demand?

Nature and Architecture

If we look at primitive architecture there is much evidence of nature incorporated in both the form and the structure of the building. Indeed we can move back from the human being into animal architecture itself and see wondrous forms and intricate skills demonstrated. But it is obvious that the incorporation of a branch or a broad palm leaf or bundle of reeds bound together to make it into a useful element is not some conscious gesture to represent nature. It is nature itself, because of the lack of anything else. It is *adaptive opportunity*.

Vernacular buildings rarely show "references" to nature in form or element, since, like the primitive shelter, they by necessity incorporate nature. Thus we see the timbers of a house with the grain and knots showing – even more impressive, the crook house, where a bent tree trunk is sliced in two to produce a pair of matching half portals. The thatch, the wattle and daub, the tiles and bricks are all from the locality and would be materials that the occupant understood – trees are felled, reeds are cut, bricks are burnt. There was no need to fashion something to look like nature; it already was nature.

By contrast, most people looking at a modern building would not know where the materials, both on the inside and out, came from or how they were produced. Even modern low-rise housing plays tricks on us – slates produced from epoxy resin, autoclaved calcium silicate bricks,

Figure 4.5 Milwaukee Art Museum, Wisconsin, US by Calatrava. Although seductively evocative of nature the form is symbolic rather than functional.

moulded doors imitating wood grain, the rooms lined with laminate floor, synthetic carpet, reconstituted stone worktops and sink, etc.

It is perhaps for this reason, our distancing from real nature, that architects and designers have become so fascinated by mimicry. Unlike the classical column this mimicry becomes symbolic by form only, not by function. Whereas the tubular steel column has a direct equivalent in the hollow stems of plants, and shares the same efficiency of material, many modern architectural manifestations of nature, as illustrated in Figure 4.5, are purely symbolic.

Implications for Environmental Design

We have suggested that man has a need for environmental stimuli and a need to respond to them. If this is true, what are the implications for the design of our buildings and cities? We have also implied that these stimuli should be due to "natural causes" and associated with the "natural" outdoors. (But this could be simply because the positive evidence available is only from cases where the stimuli are of that type.) And we have referred to this package of stimuli as *ambience*.

This prompts the following questions: is it essential to have *natural ambience* by contact with natural environmental diversity? Or can we create *artificial ambience* – where natural environmental diversity is simulated? Or even can we create *synthetic ambience* – where the diversity is artificial *and* arbitrary?

Natural Ambience

This is the conventional "adaptivist" view. The architectural interpretation is the adoption of shallow plan buildings, naturally ventilated and daylit with openable windows. Controls would be intuitive and sympathetic to occupant participation, and the spatial and technical design would provide variety and adaptive opportunity. Intermediate spaces such as atria, conservatories, loggia and verandas, free from active control, form a soft edge between the interior and exterior. Externally the architecture continues into the garden where the microclimate still shows a degree of moderation and the horticulture is applied with a range of artifice, but ultimately allows nature to dominate. The landscape design is influenced by its perception by the occupants of the building, rather than being seen as a setting for the building when viewed from outside. The principle continues at the urban scale, with accessibility of and to wildlife considered in the provision of green corridors and wild parks.

Although indoors, the occupant is placed in the natural world and the building is seen only as a mediator. And the contextual awareness does not stop at the site boundary; it is reflected in a concern for the global environment – the choice of materials and a responsible attitude to the use of energy and other resources, messages that are implied by the design of the building.

Why, then, do we have to consider the issue further?

Urban growth, the coalescing of communities, seems to be driven by a force as inevitable as the law of gravity. Unlike gravity, it is not described by a simple algorithm – rather it is the result of a complex of political, cultural, functional and environmental expedients – and, it seems, it cannot be controlled nor fully understood. The outcome, however, is clear and relevant; it has led to an ever-increasing urban density together with the increasing size of building and plan depth. This in itself removes people from the natural ambience of the world outside.

Just as at the end of the nineteenth century the developing technologies acted as a stimulus to urbanisation and the enclosure of the working environment, current technologies offer technical opportunities that creative architects find irresistible. Recent developments in materials such as glass, polymers, stainless steel, in computed structural analysis, and information technology, all facilitate the increase in size and technological complexity of the modern building. Inevitably, then, the question must arise – can we do without natural ambience? Can the environmental diversity be delivered in a different way?

Artificial and Synthetic Ambience

This is nothing new; in a less technological age evocation of the outdoors was provided by painting and sculpture, spanning perceived levels of taste from fine art to the "high-naff" of plastic flowers (with perfume!) and animated pictures of waterfalls. With current information technology it would not be difficult to offer a rich menu of naturalistic stimuli – images of landscapes and its inhabitants, sounds and even smells could be delivered deep into a building. This could transport the occupant to distant idyllic environs, or simply relay the real outdoor surroundings of the building. It could be accompanied by "naturalistic" environmental stimuli such as temperature swings and modulations of luminance and colour temperature.

Simulation and virtual reality has reached an advanced state of development – now used for applications as diverse as, for example, training in surgery, flying and presenting building "walk-throughs" from electronic moving images developed straight from CAD packages. Simulation in these circumstances is hugely successful and convincing – it is well known that airline pilots training to cope with emergencies show signs of profound stress although they are quite aware that the circumstances are not real. If this is so successful, would not the evocation of the garden outside be an easy task? But there is a difference here. In the case of the flight simulator, the illusion is the focus of interest. In contrast, the image of distant mountains projected onto the walls of a building will have to be absorbed subliminally, if it is to achieve the quality of natural ambience.

We have made the case for environmental variance and diversity in order to stimulate adaptive behaviour. But does the variance have to relate, either directly or by artificial means, to nature? Could not the thermal, visual and acoustic environment be modulated in an arbitrary way, and a new set of adaptive opportunities be created artificially? For example a temperature swing could be delivered by the air-conditioning system at the same time that a strong visual event was created by the lighting system. This could then be neutralised by an action through a graphic interface on the occupant's workstation. Would this synthetic ambience be as satisfying as walking to the window and throwing it open?

Conclusions

It appears then that our instinctive responses to the natural world are alive and well, and still make an important contribution to our health and comfort in the modern environment. However, our cultural responses have to a large part removed us from the very nature that nurtured us. There seems to be two directions to go – embrace "real nature" – naturally ventilated, daylit buildings, with user controls, set in an accessible naturalised landscape into which nature is welcomed. Or, pursue an ever more technological approach – controls with automation and IT feedback, simulation, virtual reality – a science fiction future of colour therapy rooms, sensory stimulation scenarios, and personal implants programmed to give the sensation of birdsong and spring sunshine! Indeed, this scenario has been visited by many science fiction writers, one suspects cynically, rather than enthusiastically. If successful, it would give the "advantage" of being able to completely disengage from nature – there would, for example, be no limit to the height and depth of buildings, and their occupation density.

As is customary at the end of scientific papers, we say that there is need for more research. It is hoped that this paper will help make the case for a new field of cross-disciplinary study, bringing physics, biology, psychology and sociology into the architecture and engineering of the built environment.

Acknowledgement

Based on a paper presented at the conference Eco-Architecture, held at the Wessex Institute of Technology, UK and published by WIT Press in 2006.

Notes

1. Winston, R., *Human Instinct*, Bantam Press, 2002.
2. Heschong, L., *Thermal Delight in Architecture*, Cambridge, MA: The MIT Press, 1979.
3. Schiller, G. E., A comparison between measured and predicted comfort in office buildings. *ASHRAE Transactions*, 96, 1: 609–622, 1990.
4. Fanger, P. O., How to apply models predicting thermal sensation and dis-comfort in practice, in *Thermal Comfort: Past, Present and Future*, eds. N. Oseland and M. Humphreys, Bracknell: HIS BRE Press, 1994.
5. Humphreys, M.A. and Nicol, J.F. (eds) *An Adaptive Guideline for UK Office Temperatures in Standards for Thermal Comfort – Indoor Temperature Standards for the 21st Century*, E&FN Spon, London, 1995.
6. Oseland, N.A., *A Within Groups Comparison of Predicted and Reported Thermal Sensation Votes on Climate Chambers, Offices and Homes*. Proceedings of Healthy Buildings 94, Budapest, 1994.
7. Heerwagen, J., *Adaptation and Coping: Occupant Response to Discomfort in Energy Efficient Buildings*. Proceedings ACEEE Summer Study on Energy Efficiency in Buildings, Pacific Grove, PA, 1992.
8. Bordass, W., *User and Occupant Control in Office Buildings*. Proceedings ASHRAE conference on building design, technology and occupant well-being, Brussels, 1993.
9. Baker, N., Standeven, M., Thermal comfort for free-running buildings. *Energy in Buildings* 23, 3, 1996: 175–182.
10. Guedes, M., *Thermal Comfort and Passive Cooling in Southern European Offices*. PhD Thesis, University of Cambridge, 2000.
11. Merghani, A., Exploring thermal comfort and spatial diversity, in *Environmental Diversity in Architecture*, ed. K. Steemers and M.A. Steane, London: Spon Press, 2004.
12. Nikolopoulou, M., Baker, N. and Steemers, K., Thermal comfort in outdoor urban spaces: the human parameter. *Solar Energy*, 10, 3, 2001: 227–235.
13. Parpairi, K., *Daylighting in Architecture – Quality and User Preference*. PhD Thesis, University of Cambridge, 1999.
14. Ulrich, R.S., View through a window may influence recovery from surgery. *Science*, 224, 1984.
15. Keep, P., James, R. and Inman, M., Windows in the intensive therapy unit. *Anaesthesia*, 35, 1980.
16. Robson, E.R., *School Architecture*, Leicester: Leicester University Press, 1972 (1874).
17. Heschong Mahone Group, *Daylighting in Schools*. San Francisco, CA: Pacific Gas and Electric Company, 1999.

Chapter 5

Acoustical Aesthetics, Sound and Space

Sir Harold Marshall

Introduction

Hearing is the honest sense. The eye can be tricked into believing an ephemeral surface is solid, two dimensions are three, *trompe l'oeil* in all its variety – architecture as "semblance" as Susanne Langer puts it. But sound engages only the surface impedance and thus the audible response of element or space is created – it is "honest" – what you hear is what there is.

This idea is beautifully illustrated in Figures 5.1 and 5.2.

Figure 5.1 shows the writhing forms characteristic of Zaha Hadid's architecture. This image is the staircase in the foyer of the newly opened Guangzhou Opera House. It is difficult to say by inspection which surface is solid and which is thin lining over framing. As Jonathan Glancy writes for the *Guardian*, London:[1]

> Between this exposed steel skeleton and the auditorium lie the foyers. Here, you are hard pressed to find a straight line. They waltz around the auditorium, twisting, turning, ducking and weaving. Grand stairs slope and twist majestically from the black granite floors of the foyer up to the balconies and upper tiers of the auditorium.

Architecture presenting as "semblance of solidity".

On the other hand, Figure 5.2 is the interior of the opera house. Again, Jonathan Glancy:

> The auditorium proves to be a further wonder, a great grotto like a shark's mouth set under a constellation of fairylights. The space

Figure 5.1 Guangzhou Opera House, foyer. Photo Dan Chung for the *Guardian*, London.

Figure 5.2 Guangzhou Opera House, interior. Photo Dan Chung for the *Guardian*, London.[2]

is asymmetric, but despite its unusual shape, the acoustics are perfect; the work of Harold Marshall, the veteran New Zealand acoustician. Intriguingly, he says that the strange angles of Hadid's auditorium work to produce an acoustic perfectly suited to both western and traditional Chinese opera. "There are very, very few asymmetrical auditoriums," says Marshall. "But asymmetry can be used to play with sound in very satisfying ways; it's more of a challenge tuning it, but the possibilities are greater, and this one has a beautifully balanced sound."

How it comes to be so I shall describe later, but suffice it for now to say that the impedance of every surface is known by theoretical prediction or by measurement, and by experience the acoustical designer knows how it will sound. Sound and space are one. No tricks here.

This chapter discusses why sound and space are as blest a pair of sirens as Milton's "Voice and Verse". And as he enjoined them "mix't power employ, dead things with in-breathed sense able to pierce" we may celebrate such a unity of sound and space that experience of the auditory built environment will be seen to point to the otherness beyond our metaphors.

There are of course some acoustical tricks for perception to wrestle with. The sensory recognition of the factor of two in the sound of the octave is the most significant and which in turn gives rise to the circularity of pitch and the tonal structure of Western music. Amongst the tricks is the illusion of the continuously descending tone and the cycle of fifths.

But hearing is not only the "honest sense". It is also the vigilant monitor that never sleeps, often overlooked or forgotten until a change in engine note, an unexpected click in the night, or a loaded sound such as one's name demands priority, urgency, fight or flight or some other autonomic response. For a sleeping person the transition to "wide awake" seems instant. To the yachtsman dependent on a motor to navigate a lee shore in a seaway the sound of the motor might be the sweetest note. To the hospital patient in the middle of the night and in terror of an impending procedure every click banishes the possibility of sleep. The monitor takes over.

Observation of Children

It is a long time ago that my children were toddlers but my memories of observing them in reverberant spaces remain vivid. There is nothing to suggest that they were special cases in any way other than that one's own children are always special. As I wrote 50 years ago: take a toddler into a cathedral or an empty house and observe the way the room affects the child's behaviour.[3] Somehow the empty reverberant house and the cathedral *mean* the same thing to the child. There is a hint there of how one might approach acoustical aesthetics. Or so I thought then. Recent events have led to a change in this point of view.

Learned Response

To enliven the sense of the contribution our ears make to our perception of space I liked to send my architecture students to the Auckland War Memorial Museum. This building commemorating the fallen in two World Wars stands splendidly atop one of the 50 volcanic sites in the Auckland isthmus. A neoclassical design from the 1920s it is graced with a Doric portico behind which is a grand marble entrance hall some 17 metres tall. Groups of school children and other children with their parents usually exhibit the type of behaviour I have described in my own children on entering this space, to the annoyance of the custodians. My assumption was that this dimension of architectural meaning was a given for most people – even innate.

In 2010 I had the great pleasure to meet a remarkable young man. He had been born profoundly deaf. Overcoming this handicap he went on to be head prefect in his secondary school, and now works in marketing and communications in the University of Auckland, where we met. In his early teen years he was equipped with a cochlear implant in one ear and subsequently, several years later a second device in the other. He is one of the few people I have met who can turn his hearing off completely at will. I invited him to go to the museum with his hearing off, record his impressions, then turn his hearing on again and note the differences he observed. My hypothesis was that he would notice the increased spatial sense with the hearing on. It will be best if I allow his report to speak for itself. Here it is in full:

Verbatim Report

I went to the museum on Saturday morning and it was a very interesting experience! I switched off my cochlear implants as I walked through the main entrance, then switched them back on later.

Here are my observations:

- Without sound, the foyer seemed empty and expansive – I felt a bit lost. When I switched my cochlear implants on

Figure 5.3 Auckland War Memorial Museum. (Source: SimonHo.org.)

later, it suddenly seemed busy and chaotic with a humming noise in the background and people chattering as they arrived through the entrance.

- In the silence I also noticed that I was a lot calmer, focusing on exhibits and soaking up details.
- As soon as I could hear, I became more rushed and was distracted by noises so I spent time trying to locate the sources of sound (often a child calling out, footsteps, background music, voiceovers on video screens). I found it harder to absorb details from the exhibits I was looking at.
- There was a tree-house playground in the Maori Court, surrounded by exhibits and low walls. I found this was the loudest room in the museum! Children were playing on the playground and their shouts were echoing endlessly around the walls of the small room – quite a noisy experience!
- I noticed little children were hyperactive in rooms that echoed with low roofs, but they seemed quieter (perhaps over-awed) in wide open spaces. There was a family group in the Kai to Pie food exhibition, which was darkened with a very low roof, giving the effect of being in a tunnel. This group had at least five young children who ran amok in the exhibition squealing and shouting, but the kids went quiet as soon as they walked out into the main atrium.
- Later on, a band started playing in the atrium for South Africa Day and I could hear very clearly because there was a very high roof with a round wooden structure above the stage.
- I noticed there was a lot more ambient/background noise in the upper floors. The wooden floors made footsteps extra loud, especially in the corridors through Scars of the Heart.
- I also noticed a lot of noise carrying up to the top floor from the foyer below (you can look down to the foyer over some railings).
- Overall, I found it easier to concentrate and relax in wide open spaces than in enclosed spaces that seemed to amplify sounds.

Hope this helps – it was good being able to visit the museum on the weekend!

Comment

This fascinating account sheds great light on a number of aspects of my hypothesis. The expected enhancement of spatial sense did not occur, the children were more hyperactive in the lower ceiling spaces suggesting that the reverberant level and small room effects may be more significant than the reverberation time. Most importantly, the effect of the reverberant sound was simply noise to him. Though the sample of one is so small, I concluded from his account that response to the acoustical dimension of architectural space is learned in the early years of one's life. During those years he had no hearing and so never learned either in absolute terms nor in culturally determined terms the response such spaces as memorials or cathedrals might be expected to produce. When children become rowdy in such spaces, they are in fact *learning* in an environment that is new to them.

Another point that struck me is the importance of localisation to him in trying to deal with the distracting noises. Had he been listening with only one side I expect the problem would have been much worse as users of hearing aids would confirm. But this fact also has relevance for involvement in musical performances especially of ensembles. Reverberance and the presence of early lateral reflected sound, both preferred for listening to music of this type because of the enveloping sound, both weaken the localisation of the source. A source readily localised can also be adapted to as he was seeking to do – the antithesis of desired music-listening conditions.

For reasons of privacy this young man cannot be named but I am profoundly grateful to him for taking the time to fulfil my request so expertly.

Some Acoustical Realities

What then are the acoustical properties of enclosure, which are readily perceptible and so might contribute to space meaning? The primary attribute is of course reverberation, then in large spaces and/or with energetic directional sources, and/or focusing surfaces, echo. Other anomalies such as flutter, room resonances and whispering galleries, while fun to listen to, are dependent on specific geometries and so have less general applicability to architecture.

Echoes, Reflections and Reverberation

How do "reflections" and "echoes" differ from reverberation? All are generated by the reflection of sound from surfaces and there will of course be overlap. But there is a simple perceptual test to distinguish between them for the sake of this discussion. An echo is a reflection that can be *located* as a discrete acoustical event, distinct in space and/or time from the source that generated it. The early reflections are *measurably* discrete but perceptually merged with the source and contribute to its loudness, clarity and presence. They are perceptually integrated with it. The time over which this integration occurs is determined by the signal. For Western speech it is around 50 milliseconds, for music it is very dependent on the genre, and temporal structure. The remainder of reflections, later and declining in strength as their number increases, merge with one another, by and large without a specific direction, to form the reverberation. The shape of the reverberant field as it decays must be considered in the shaping of the room.

Impulse Response

It is a happy fact that the response of a room (or any linear system) to an input signal is fully characterised by its so-called *impulse response*. This theoretical construct is approximated by a hand clap, a spark or a shot at room scale or in models, and a great deal can be learned from this type of excitation. Capture the impulse response monaurally and most of the measures derived for speech or musical communication in rooms during the past half century can be calculated from it. Capture it binaurally in a dummy head and the remainder of measures perceptually important in concert halls are available.

But even without instrumentation there is much that can be gleaned from a hand clap. Indeed the expert listener can detect the size of the space, and the predominant construction materials in the surfaces and even the reverberation time with remarkable accuracy. Solid stone or masonry rooms tend to be boomy. Framed constructions with relatively thin linings tend to support higher frequencies. In the presence of "fuzz" such as carpet and drapes in such rooms there is a peculiarity in the reverberation. Low frequencies are absorbed by the wall linings, high by the fuzz. That leaves the octave around middle C (c250 Hz) excessively bright and loud – sufficiently so to make teaching music in such a room very difficult.

Architectural Significance of Reverberation

The rooms in which reverberation is most pronounced in the pre-industrial age tend to be places of high symbolic meaning whether of religious or memorial significance. This is because often these were large spaces needed for ceremonial occasions to accommodate many people, and because these were constructed at great cost to their communities in permanent materials – typically stone or other types of masonry with low acoustical absorption, wherever self-awareness in the context of symbols is sought. (It is no coincidence that the voice of the gods is always reverberated in film, radio or TV whether in *Buffy the Vampire Slayer*, *The Lord of the Rings* or detergent ads.) These are of course modern degeneracies of what appears to be a very long history.

The reverberation in these rooms enhances personal identification with the symbols they embody.

Prehistory

Enclosure was a special experience to our forebears in the open, too, 25,000 years and more ago. The only reverberant spaces they encountered were the caves and grottos and as a special case forests. These must very soon have taken on mystical significance. There can be little doubt that in

Figure 5.4 The Waitomo Caves Choir in the "Cathedral Cave".

the caves of France, Lascaux and others the wonderful and moving paintings of the animals with which early humans shared the planet had a religious significance. There is a proposal that the musical instruments such as flutes found in the caves were used to excite the natural resonances and that where these had maxima were regions of greatest concentration of images.[4]

Little has changed. Such spaces in the twenty-first century are often claimed to have "perfect acoustics" and in NZ in its most famous limestone caves a resident choir gives regular concerts there because of this fact.

What then of forest reverberation? This differs from reverberation in a room in several ways. In an important study a direct comparison was made between the two reverberant fields – in a forest and in a hall.[5, 6] It is clear from these studies that reverberation time in a forest is dependent on distance from the source where in a normal room the reverberation time approximates the value predicted by the Sabine equation irrespective of location. For the purposes of this argument it is sufficient to assert that only in forests or in grottos would reverberation have been encountered. Could that have been the origin of the "sacred groves" of many Mediterranean cults? Or in England the imperative for the creation of the timber circles such as that recently reconstructed at Durrington Walls? It is certain that such an array would have generated significant reverberant fields.[7]

Establishment of Priority for Acoustics – Importance of Listening

It is clear that the preoccupations of architects are with building form. Yet the form is crucially important to the sound that will result. How then can the acoustician – assuming one is appointed at this stage – establish priority for acoustical consideration? *Listening* is the key.

Figure 5.5 The Durrington Walls reconstruction of the timber circles.

One anecdote illustrates ways in which this happens in a process I consider near ideal and centred on the form of the room.

The Orange County Centre for the Performing Arts 1980–86 – now Segerstrom Hall – remains for me the ideal process for the design of a concert hall. In large measure that is because of the "architecture by team" approach led by Charles Lawrence of CRS, Houston, Texas. A

Figure 5.6 Segerstrom Hall, Orange County.

rigorous programming session (Problem Seeking) leads on to the design squatters. The programming had resulted in a broad fan shape before there was acoustical design input, a shape anathema to the desire for strong lateral sound! How to achieve this within the fan-shaped envelope was the first task of the squatters. After the entire team had listened to head oriented stereophonic recordings matched to slides of the relevant rooms the entire team sought a room form within which the lateral sound would occur at every seat. The result is history. That it worked so well in this case is due in large measure to the inter-personal expertise of CRS and their commitment to the philosophy of "Architecture by Team". It is essential in their view that the client/owner and all other consultants be included in the team. In the event, the architectural solution to the problem was proposed by the acousticians, not in opposition to the other team members but in an atmosphere of excitement and consensus.

Project managers who intervene between owner and designer, theatre consultants or lighting consultants with a single fixed idea for the way a space must be, interior designers without regard for anything other than their own aesthetic scheme are all fatal to the team design process advocated here.

Acoustical Design

Of course as an acoustical designer, design process has held a profound interest for me. How does one encourage architectural creativity in the context of a concert hall, a church or other communications space, or indeed a multiple-use space, without infringing the "laws" of room acoustics? The answer to this conundrum lies in the fact that the so-called laws are essentially limited even in their own terms and if one examines their experimental derivation the picture is even less convincing. For some years I have advocated that the various room acoustic metrics should be taken merely as evidence that gross errors have been avoided and that communication between "a creative and receptive architect and an acoustician able to communicate at the level of the architectural intention" is essential.

If this does not happen and the acoustician is limited to the acoustical engineering the conflict spelled out by the late Jean Paul Vian is probably inevitable.[8]

> When designing a new opera or concert hall, the architectural requirements for excellence are most of the times conflicting with the similar acoustic requirements for excellence. Good compromises have to be sought; with the well known difficulty that seeing mostly predominates on hearing.

Let me now suggest a way in which the term, so beloved of engineers – "compromise" – is unnecessary.

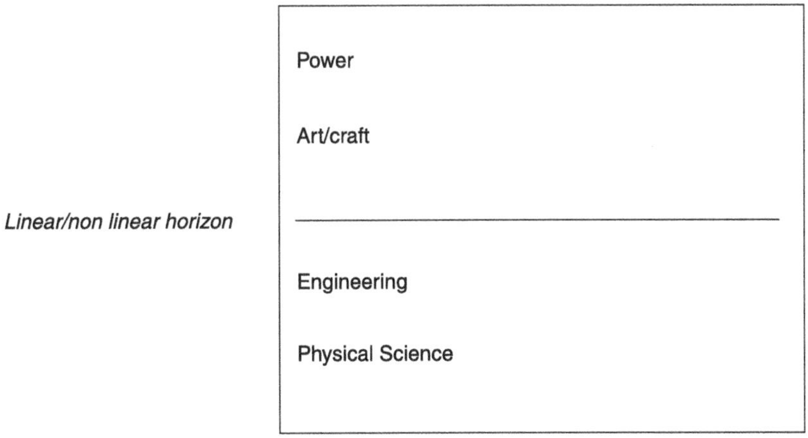

Power

Art/craft

Linear/non linear horizon

_____

Engineering

Physical Science

Figure 5.7 Hierarchy of knowledge and skill in concert halls.

Comment: Hierarchy of Knowledge about Concert Hall Design

I refer here to a suggestion from my 1989 plenary lecture at the Belgrade ICA and which I have had no occasion in the subsequent years to modify. After identifying four levels of knowledge about concert hall design I pointed out (heresy at that time) that physics is necessary but insufficient for a successful design. Broadly speaking there is a horizon separating physical science and its derivatives (such as engineering) that succeeds at a linear and reductive level, from art/craft (design) and political power, which are notoriously non-linear and holistic. Nothing in the latter two is permitted to contradict the former but neither are art/craft skills or political power comprehensible solely in terms of physical science. Such considerations underlie the topic addressed here.[9]

To progress from the levels of physical science and engineering to acoustical design, it is necessary to engage with meaning in architecture and music.

Meaning in Architecture

There are innumerable examples of architectural writing on this topic. I have chosen one author – Daniel Libeskind.

> These programmatic activities are given three-dimensional depth, not in neutral containers, but in functional and emblematic spaces, each of which has a density, materiality, temperature, acoustical quality, atmosphere and gravity which are not fully accessible to the abstraction of words, but rather to concretely embodied experience.[10]

Or as Thomas Aquinas (1224–1274) wrote 800 years ago: ". . . *quia sensus delectatur in rebus debite proportionatis, sicut in sibi similibus; nam et sensus ratio quaedam est, et omnis virtus cognoscitiva*" ["for the senses delight in things duly proportioned as in something akin to them; for the sense too is a kind of reason as is every cognitive power"].[11]

Meaning in architecture is in fact closely analogous to meaning in music – multilayered, literal or figurative and "not fully accessible to the abstraction of words". The best examples I can think of are the Cantatas of J.S. Bach. There *is* a literal function, Bach's "sermons", supported by musical metaphor, rigorous abstract order and transcendent, even sublime creativity.

Recognition of the dimension of meaning in both architecture and music has to be the starting point for design in communication spaces. It is essential for the designer to grasp this fact and rhetorical theory may be helpful in this.

Rhetoric

In this section I am drawing on the writings of Suzanne Langer and a recent paper in the *American Communication Journal* by Jon Radwan: "Mediated Rhetoric: Presentational Symbolism and Non-Negation".[12]

Suzanne Langer was writing philosophy about 60 years ago but has had a resurgence in relevance in the new Millennium as the blandishments of the quasi-liberation of postmodernism have faded. But first, what is rhetoric? Radwan has it thus: "*the process of mutual influence that characterizes social exchange*", in other words communication. "It is the basic human process in the sense that we are social animals who look to one another for cues." Meaning is the product of these cues and amongst the media in which they occur are space and sound.

So how does rhetoric help us? It comes as no surprise to learn that Kenneth Burke[13] introduces his *Rhetoric of Motives* with a discussion of persuasion through language.

"All told, persuasion ranges from the bluntest quest for advantage . . . as in propaganda . . . to a pure form that delights in the process of appeal for itself alone, without any ulterior motive" and "through all the mysteries of social status to the mystic's devout identification with the source of all being". The quotes are from Note 4.

There is indeed an acoustical dimension to such an epiphany that is not able to be captured verbally as we shall see; it is essentially experiential.

It is more of a surprise to discover the presence of the pervasive negative in most *verbal* discourse. Discursive meaning arises through structural negation and difference. This is the stuff of the linear and reductive characteristic of the hierarchical elements below the horizon in Figure 5.3. Langer on the other hand is very clear that communication of human symbols may be discursive (as in the building blocks of rhetoric) but that communication of symbols through art forms is different in kind. It is *presentational*. That is, its meaning occurs in the fact that the observer/auditor and the work are *present* to each other. The meaning is only accessible *in experience* as Libeskind wrote. Langer goes on to discuss painting,

Figure 5.8 The competition design for La Philharmonie de Paris.

architecture and music in these terms in a way that makes great sense in terms of acoustical design.

Shared meaning is the site for a meeting of minds.

The acoustician must seek such a meeting of minds with the architects. From long experience we expect architects whose skills can obtain commissions as significant as opera houses and concert halls to be receptive to the underlying technology and excited by it. Our aim is a design milieu in which *sound, space and performance* are one even though it is generally not necessary to be explicit about this. Presentational communication is the aim.

So, where do the architect's priorities lie? There are many ways to describe the imperatives that motivate architects. The one that I like is that architectural design turns on a sense of a building *wanting to be*. This embraces both the functional and the emblematic. Of course, I borrowed that from Louis Kahn from Princeton University over 40 years ago.

Acoustical Design Values; Paris

The question is, can acousticians have the confidence to think like that – that what "wants to be" in the context of a new building form includes the sound in the space. *That* is acoustical design. It is a skill that I have actively sought, I confess, since 1962 when I first published a paper in the *Architectural Science Review*,[14] which described that quest. Sound and space, I believe, can be pre-experienced in the designer's mind, just as the architect pre-experiences the space, the visual space. An example of the needed meeting of minds is the winning design for the Philharmonie de Paris, with Atelier Jean Nouvel and Marshall Day Acoustics.

Guangzhou

Now we can comprehend also how the Guangzhou Opera interior came into being. In the words of Simon Yu the resident associate for Zaha Hadid, largely responsible for realizing this great building:

> Reference was made to Cardiff and more emphasis on the asymmetry as we progressed – to our delight and excitement when we engaged Harold (I remember it so well) that he was "one mind" with the architects and so began this labour of love . . .![15]

We are glad that there was a meeting of minds between the architects and ourselves. You cannot look at the space in Figure 5.3 and say "the acoustics begins here . . . or there". Within that meeting we have been able to predict the physical metrics of sound in rooms to ensure that the objective design guides are fulfilled. We value asymmetry – orchestras, after all, are very asymmetrical in their radiation of sound. In fact I too recall our first discussion about asymmetry and suggesting the reflecting surfaces from left and right as arms embracing the stage, within the established architectural idiom. This became the defining character of the space in the architects' hands.

So where in this scheme of things do the conventional metrics for room acoustics fit?

Conventional Metrics

As responsible consultants we use all the resources that are available to verify and refine the acoustical designs made. It is absolutely imperative that a new design is checked with every available metric that is supported by research. This may include one-tenth scale or one-twenty-fifth scale models with dry air or numerical compensation via MIDAS. We as acousticians provided the motivation for the development of the MIDAS software for digital data acquisition at all scales, and most recently we push ODEON to its very real limits for rooms of great complexity such as Guangzhou in prediction of these metrics. Auralisation has its place here. All of these are intended to prove the design. Our aim is a room in which sound and sight together contribute to the concert experience. The metrics prove the design, they are not the design themselves, but at the very least they are a check that gross errors have been avoided. They also enable the results to be refined and allow us to present them with confidence to the client. The variety of spaces that we have produced is evidence that we do not subscribe to a recipe for the ideal concert hall.

Having a Cuppa

> Aspirin, Surgery, And a cup of tea,
> but the greatest of these is a cup of tea.
>
> "Communion" by Glenn Colquhoun

To come down to earth . . .

The simplest communication environment – a conversation over a cup of tea – may be compromised by the acoustical conditions prevailing.

Yet it is in such communication that we engage with one another in the "basic act which defines our humanity".

When churches seek acoustical advice on the design of worship and meeting spaces the acoustical design of the principal space is straightforward but often (possibly due to budget constraints) the adjacent foyer is

Figure 5.9 King's College Chapel, Cambridge, England.

neglected. The unpleasant conditions in such spaces – often used for morning tea or gatherings – can overshadow the acoustical design work in the adjacent room.

Such a community in NZ's Southland province was deeply disappointed in the gathering space associated with their refurbished church and asked for my help. Fortunately I was in Southland visiting family only a hundred kilometres from the town. I visited, estimated the RT and analysed the problem. The technical argument is found in Appendix 1.

Because of happy numerical coincidences the equation relating the number of people (N), the metrics of speech communication, the absorption inherent in a space (A_0) and the additional absorption needed (A_1) to ensure adequate communication between any two of them simplifies, in general, to:

$$A_1 = 1.5N\text{-}A_0$$

Where A_0 is given by $A_0 = 0.161V/T$, V is room volume m^3 and T is the design RT

In typical spaces, A_0 is likely to be close to one-third of the required total and the area of installed absorbent required becomes equal to the number of people present at the rate of 1 square metre per person. This elegant result takes into account all the factors, including the frequencies in which the vocal power resides, the signal-to-noise ratio for adequate communication in noise, speaker directivity, absorption per person (standing, in an overcoat) and a person-to-person distance of 0.75 metres. "One square metre per person" should be within the numeracy skills of all!

Conclusions

I started this chapter with the hope that "we may celebrate such a unity of sound and space that experience of the auditory built environment will be seen to point to the otherness beyond our metaphors". By considering the presentational rhetoric of architecture and music one comes to an understanding of meaning in both these art forms, the necessity for a meeting of minds in acoustical design, and how indeed transcendence blossoms in the union of sound and space.

Appendix 1

An article by Marshall Long from 2005 (*Acoustics Today*, Volume 1, Number 1) suggested a rather different approach from the RT approach I have usually used in such rooms. He adopts a signal/noise basis for predicting the amount of absorption necessary in a restaurant to ensure conversation is possible across a table but incomprehensible at the next table. S/N −6dB is the criterion for 60 per cent of sentences correctly identified. This is equated with the difference between the direct sound on axis of a talker at

1 metre and the reverberant noise from N other persons. I have adapted this idea to the gathering space in a church foyer in Southland.

My analysis of this simple problem is as follows:

Room Volume = 435 m^3, RT unocc. 1.2 sec (estimated on site)

Total absorption room unoccupied, A_0 = 58m^2

Number of people present N

Number talking N/2

Power Level of talker L_w = 70dB

Directivity Q = 2

Distance of receiver = 0.75m

Absorption per person standing adult in overcoat at 250Hz, $s\alpha$ = 0.5m^2 (approximated from Kath and Kuhl – quoted in Rossell and Vincent, Proc. International Symposium on Room Acoustics, ISRA, Seville, 2007 (and in *Acustica* 15 [1965]: 127–131).

Reverberant Level in unocc. rm L_{pr} = L_w+10log[4/A_0] + K (K is negligible with A in m^2)

With N persons present L_{pr} = L_w + 10log (N/2)+10log[4/(A_0+Nsα)]

With N persons and A_1m^2 installed absorption
= L_w + 10log (N/2) + 10log[4/(A_0 + Nsα + A_1)] α

SPL of the direct sound @ 0.75m L_{pd} = L_w+10 log[Q/4πr^2]+ K = 64dB

Evaluating these expressions gives:

N	0	10	20	40	100
L_{pd} – L_{pr} as built	+6dB	−2.8	−3.7	−5.5	−8.7
L_{pd} – L_{pr} including 65m^2 absorbent A_1		+1.8	−0.9	−2.6	−7

That "as built" the foyer is marginal for 40 people, entirely unsatisfactory for 100.

Figure 5.10 The installed absorbent ceiling in the gathering space.

In this case, 65 metres squared for practical reasons seemed the maximum possible for the installed absorbent. It is chosen to have a peak absorption of about 100 per cent at 250 Hz. In a design rather than a remedial situation as this was, one would solve the signal/noise equation for the appropriate area A_1 thus:

$$L_{pd} - L_{pr} = 64 - 70 - 10\log(N/2) - 10\log[4/(A_0 + Ns\alpha + A_1)]$$
$$= -6\text{dB} \tag{1}$$

$$-6 - 10\log(50) - 10\log[4/(58 + 100*0.5 + A_1)] = -6\text{dB}$$

$$\log[(108 + A_1)/4] = 1.7$$

$$108 + A_1 = 200$$

$$A_1 = 92\text{m}^2$$

Alternatively one could solve for N, given 65m^2 for A_1

$$-6 - 10\log(N/2) - 10\log[4/(58 + 0.5N + 65)] = -6$$

$$(58 + 65) = 3N/2$$

$$N = 246/3 = 82 \text{ persons}$$

In this room there will be a residual deficit for gatherings over 82 people.

The church immediately instructed their architect to implement this change with the result that the problem disappeared.

Notes

1. www.guardian.co.uk/artanddesign/2011/feb/28/guangzhou-opera-house-zaha-hadid (accessed 4 February 2013).
2. www.guardian.co.uk/artanddesign/gallery/2011/mar/01/zaha-hadid-guangzhou-opera-house-in-pictures (accessed 4 February 2013).
3. Marshall, A.H., "An Acoustical Dimension of Space Meaning", *J. Architectural Assn.* (May 1968).
4. Dauvois, M. and Boutillon, X., *Études Acoustique au Réseau Clastres: Salle des peinture et lithophone naturels*, Bulletin de la Société Préhistorique Ariege-Pyrénées (1990).
5. Hiroki Sakai, Shin-ichi Sato and Yoichi Ando, "Othogonal factors in Sound fields in a forest compared to those in a concert hall", *J Acoust Soc Am.* 104(03) (1998): 1491–1497.
6. Wiens, T., Bradley, S. and George, K., "Experimental characterization of sound propagation in a dense NZ Forest", *Proc Internoise* 998 (2008).
7. www.channel4.com/history/microsites/T/timeteam/2005_durr.html (accessed 4 February 2013).
8. www.researchgate.net/publication/228473839_Acoustic_Design_of_the_National_Grand_Theatre_of_China_an_attempt_to_get_the_rich_sound_of_a_modern_opera_in_a_classical_horse_shoe_theatre (accessed 18 February 2013).
9. Marshall, A.H., "Recent developments in acoustical design process", *Applied Acoustics*, 31(7) (1990): 7–28.
10. Libeskind, D., *New Statesman*, 24 June 2002, p. 36.
11. Summae Theologiae Angelici Doctoris Sancti Thomae Aquinatis, Prima Pars, Quaestio V, De Bono In Communi In Sex Articulos Divisa Re: ST I, Q.5 A.5.
12. Jon Radwan, "Mediated Rhetoric: Presentational Symbolism and Non-Negation", *Am. Com. Jnl* 5(1) (2001): 1–12.
13. Burke, K., *A Rhetoric of Motives*, Berkeley, University of California Press (1969).
14. Marshall, A.H., "Acoustics Courses in a School of Architecture", *Arch. Science Review* (November 1962).
15. Simon, K.M. Y., Letter to Peter Exton, 24 March 2011.

Chapter 6

Existential Comfort

Lived Space and Architecture

Juhani Pallasmaa

Introduction: Settling the Mind

The task of architecture is usually seen in terms of functional performance, physical comfort and aesthetic values. Yet, its role extends beyond the material, physical and measurable conditions into the mental and existential sphere of life. Buildings do not merely provide physical shelter and protection; they also mediate mentally between the world and the human consciousness. As the French philosopher Gaston Bachelard appropriately states: "[The house] is an instrument with which to confront the cosmos."[1] In addition to housing our fragile bodies, architecture settles our restless minds, memories and dreams. In short, architectural constructions organize and structure our experiences of the world; they project distinct frames of perception and experience, and provide specific horizons of understanding and meaning. Man-made structures also concretize the passage of time, represent cultural hierarchies and give a visible presence to human institutions.

       The analysis of the mental task of architecture takes us outside of physics and physiology, and even beyond psychology, into our unconscious motifs and memories, desires and fears. The mental sphere of architecture cannot be approached by instruments of measurement; the poetic essence of architecture is grasped through an embodied encounter, intuition and empathy. Architecture focuses on lived experiential essences and mental meanings; this very aim also defines the architect's true approach and method. As Jean-Paul Sartre states: "Essences and facts are incommensurable, and one who begins his inquiry with facts will never arrive at essences. [. . .] Understanding is not a quality coming to human reality from the outside, it is its characteristic way of existing."[2] All artistic works, including architecture, seek this natural mode of understanding that is entwined with the very act of being. Consequently, the true essence of architecture does not arise from an aesthetic aspiration; it originates in an existential desire.

Jorge Luis Borges describes memorably the essence of the poetic experience: "The taste of the apple [. . .] lies in the contact of the fruit with the palate, not in the fruit itself; in a similar way [. . .] poetry lies in the meeting of poem and reader, not in the lines of symbols printed on the pages of a book. What is essential is the aesthetic act, the thrill, the almost physical emotion that comes with each reading."[3]

Similarly, the meaning of architecture actualizes in the unique encounter of space and the person, in the very merging of the world and the dweller's sense of self. In the poetic survey of architecture as well as poetry,

Figure 6.1 Architecture and cosmic dimensions – Arrival Plaza at the Cranbrook Academy (Bloomfield Hils, Michigan, 1994) terminates the entry road to the campus, creates a sculptural focus and mediates the viewer's awareness with celestial bodies and time.
Source: Photo Balthazar Korab.

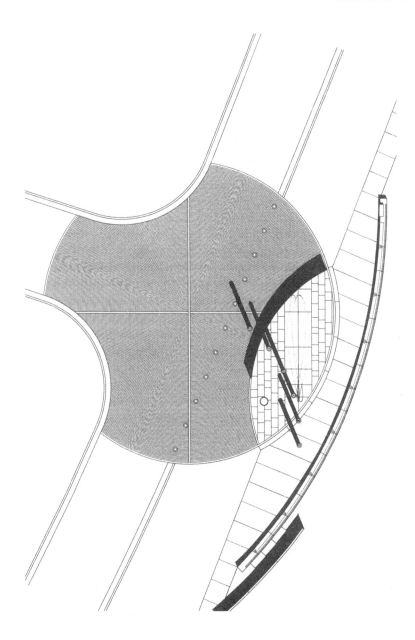

Figure 6.2 Site plan of the Arrival Plaza and Cosmic Instrument.

the perceiving and experiencing self, the first person, has to be placed in the center.

The mental dimension of building is not a mere surplus value of utility and reason; it is its very essence. Soulless buildings are detrimental regardless of their functional, thermal, ergonomic, acoustic, etc., qualities, because they fail to root us in our lived reality and to mediate between the world and our consciousness. They do not enable us to understand ourselves. My essay attempts to identify the existential, sensory and embodied ground of architecture that gives rise to *mental and existential comfort.*

Beyond Vision

The art of architecture continues to be regarded, theorized and taught as an art form of the eye. As a consequence, it is dominated by considerations of retinal qualities, such as visual composition, harmony and proportion. The dominance of vision has never been stronger than in our current era of visual images and their industrial mass production, "an unending rainfall of images," as Italo Calvino puts it.[4] As a consequence of this biased emphasis, buildings are turning into objects of momentary visual seduction, and they are losing their material presence, plasticity and hapticity. Buildings have become aestheticized objects that are externally viewed rather than being lived as inseparable parts of our very awareness and sense of life. Buildings are increasingly objects of admiration instead of being "instruments to confront the cosmos." This unfortunate development is strengthened by the development of building processes, techniques and materials towards uniformity and a sense of immateriality as well as today's obsessive objectives of economy and instant gratification.

The sensory and mental impoverishment of contemporary retinally biased environments has made it clear that profound architecture is a multi-sensory and existential art form; our buildings need to address our senses of hearing, touch, smell and even taste, as much as pleasing the eye. They need to provide us our corner in the world, instead of mere visual titillation. Maurice Merleau-Ponty argues strongly for the integration of the senses:

> My perception is [therefore] not a sum of visual, tactile, and audible givens: I perceive in a total way with my whole being: I grasp a unique structure of the thing, a unique way of being, which speaks to all my senses at once.[5]

The true wonder of our perception of the world is its very completeness, continuity and constancy regardless of the fragmentary nature of our observations. Meaningful architecture facilitates and supports this experience of completeness.

The loss of hapticity, sense of intimacy and nearness has particularly negative consequences as it evokes experiences of alienation, detachment and distance. These experiences give rise to the feeling of "existential outsideness," to use a notion of Edward Relph.[6] In order to root us in our world, buildings need to go beyond sensory comfort and pleasure into the very enigma of human existence. "Writing is literally an existential process," Joseph Brodsky argues, and the same must be said of both the conception and experience of architecture.[7]

Primacy of Touch: Hapticity of the Self-Image

All the senses, including vision, are extensions of the tactile sense; the senses are specializations of skin tissue, and all sensory experiences are modes of touching, and thus related with tactility. "Through vision we touch the sun and the stars," as Martin Jay poetically remarks in reference to Merleau-Ponty.[8] Our contact with the world takes place at the boundary line of the self through specialized parts of our enveloping membrane.

The view of Ashley Montagu, the anthropologist, based on medical evidence, confirms the primacy of the haptic realm:

> [The skin] is the oldest and the most sensitive of our organs, our first medium of communication, and our most efficient protector [. . .] Even the transparent cornea of the eye is overlain by a layer of modified skin [. . .] Touch is the parent of our eyes, ears, nose, and mouth. It is the sense, which became differentiated into the others, a fact that seems to be recognized in the age-old evaluation of touch as "the mother of the senses."[9]

In their book *Body, Memory and Architecture*, an early study in the embodied essence of architectural experience, Kent C. Bloomer and Charles Moore emphasize the primacy of the haptic realm similarly: "The body image [. . .] is informed fundamentally from haptic and orienting experiences early in life. Our visual images are developed later on, and depend for their meaning on primal experiences that were acquired haptically."[10]

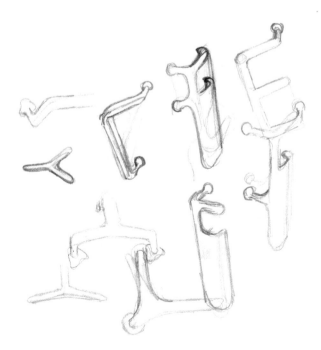

Figure 6.3 Design for the tactile sense. Sketches for door pulls; page of a sketch book, early 1980s. Source: Author.

Touch is the sensory mode that integrates our experiences of the world and of ourselves. Even visual perceptions are fused and integrated into the haptic continuum of the self; my body remembers who I am and how I am located in the world.

In Marcel Proust's *In Search of Lost Time: Swann's Way*, the protagonist waking up in his bed reconstructs his identity and location through his body memory:

> My body, still too heavy with sleep to move, would endeavor to construe from the pattern of its tiredness the position of its various limbs, in order to deduce therefrom the direction of the wall, the location of the furniture, to piece together and give a name to the house in which it lay. Its memory, the composite memory of its ribs, its knees, its shoulder-blades, offered it a whole series of rooms in which it had at one time or another slept, while the unseen walls, shifting and adapting themselves to the shape of each successive room that it remembered, whirled round it in the dark. [. . .] my body, would recall from

Figure 6.4 Design for the human body, laminated plywood and carbon fiber, chromed spring steel base, chair, prototype, 1991. Photo Rauno Träskelin.

each room in succession the style of the bed, the position of the doors, the angle at which the sunlight came in at the windows, whether there was a passage outside, what I had had in mind when I went to sleep and found there when I awoke.[11]

The writer's description provides a powerful example of the intertwining of the body, memory and space. My body is truly the navel of my world, not in the sense of the viewing point of a central perspective, but as the sole locus of integration, reference, memory and imagination.

The Unconscious Touch

We are not usually aware that an unconscious experience of touch is unavoidably concealed in vision. As we look, the eye touches, and before we even see an object, we have already touched it and judged its weight, temperature and surface texture. Touch is the unconsciousness of vision, and this hidden tactile experience determines the sensuous qualities of the perceived object. The unconscious sense of touch mediates messages of invitation or rejection, nearness or distance, pleasure or repulsion. It is exactly this unconscious dimension of touch in vision that is disastrously neglected in today's retinal hard-edge architecture. This architecture may entice and amuse the eye, but it does not provide a domicile for our bodies and minds.

The haptic fusion with space and place surpasses the need for physical comfort and the mere desire to touch. Bachelard recognizes the desire for a total merging of the self and the house through a bodily intertwining as he writes:

> Indeed, in our houses we have nooks and corners in which we like to curl up comfortably. To curl up belongs to the phenomenology of the verb to inhabit, and only those who have learned to do so can inhabit with intensity.[12]

The pleasure of curling up also suggests an unconscious association between the images of room and the womb; a protective and pleasurable room is a constructed womb, in which we can re-experience the undifferentiated world of the child, the forgotten infant concealed in our adult bodies. Today's architecture, however, tends to offer us mere wombs of glass to inhabit.

Merleau-Ponty writes emphatically:

> We see the depth, speed, softness and hardness of objects – Cézanne says that we see even their odour. If a painter wishes to express the world, his system of colour must generate this indivisible complex of impressions, otherwise his painting only hints at possibilities without producing the unity, presence and unsurpassable diversity that governs the experience and which is the definition of reality for us.[13]

In developing further Goethe's notion of "life-enhancing art" in the 1890s, Bernard Berenson suggested that when experiencing an artistic work we actually imagine a genuine physical encounter through "ideated sensations." The most important of these ideated sensations Berenson called "tactile values."[14] In his view, the work of authentic art stimulates our ideated sensations of touch, and this stimulation is life enhancing.

In my view, also, a profound architectural work generates similarly an indivisible complex of impressions, or ideated sensations, such as experiences of movement, weight, tension, texture, light, color, formal counterpoint and rhythm, and they become the measure of the real for us. Even more importantly, they become unconscious extensions of our body and consciousness. When entering the courtyard of the Salk Institute in La Jolla, California, a couple of decades ago, I felt compelled to walk directly to the nearest concrete surface and sense its temperature; the enticement of a silken skin suggested by this concrete material was overpowering. Louis Kahn, the architect of this masterpiece, actually sought the gray softness of "the wings of a moth"[15] and added volcanic ash to the concrete mix in order to achieve this extraordinary matte softness.

True architectural quality is manifested in the fullness, freshness and unquestioned prestige of the experience. A complete resonance and interaction takes place between the space and the experiencing person. This is the "aura" of a work of art observed by Walter Benjamin.[16]

Space and the Self

Our normal understanding of space, the commonplace "naïve realism," regards space as a measureless, infinite and homogenous emptiness in which objects and physical events take place. Space itself is seen as a meaningless continuum; signification is assumed to lie solely in the objects and events occupying space. The assumption that environment and space are neutral concepts existing outside man, thus continues to be axiomatic in everyday life. Yet, it is precisely this separation of man and environment that anthropologist Edward T. Hall views as one of the most destructive unconscious cornerstones of Western thinking.[17]

One of the influential thinkers to point out the essential existential connection between space and the human condition, the world and the mind, was Martin Heidegger. He paid attention to the connectedness – or perhaps we should say, the unity – between the acts of building, dwelling and thinking. He links space indivisibly with the human condition:

> When we speak of man and space, it sounds as though man stood on one side, space on the other. Yet space is not something that faces man. It is neither an external object nor an inner experience. It is not that there are men, and over and above them space [. . .][18]

Surely, we do not exist or dwell detached from space, or in an abstract and valueless space; we always occupy distinct settings and places that are intertwined with our very consciousness. Lived space always possesses specific characteristics and meanings. Space is not inactive; space either empowers or weakens, charges or discharges. It has the capacity to unify or isolate, embrace or alienate, protect or threaten, liberate or imprison. Space is either benevolent or malicious in relation to human existence.

The world around us is always organized and structured around distinct foci, such as concepts and experiences of homeland, domicile, place, home and self. Also our specific intentions organize space and project specific meanings upon it. Even one's mother tongue, such as the unconscious notions of above and below, in front and behind, before and after, affect our understanding and utilization of space in specific and preconditioned ways. Also space and language are intertwined. As we settle in a space, it is grasped as a distinct place. In fact, the act of dwelling is fundamentally an exchange; I settle in the place and the place settles in me. This merging of space and self is one of the founding ideas of Maurice Merleau-Ponty's philosophy that offers a fertile conceptual ground for the understanding of artistic, architectural and existential phenomena.

Existential Space

We do not live in an objective world of matter and facts, as commonplace naïve realism assumes. The characteristically human mode of existence takes place in worlds of possibilities, molded by our capacities of memory, fantasy and imagination. We live in mental worlds, in which the material and the spiritual, as well as the experienced, remembered and imagined, fuse completely into each other. As a consequence, the lived reality does not follow the rules of space and time, as defined and measured by the science of physics. We could say that the lived world is fundamentally "unscientific," when assessed by the criteria of empirical science. In its diffuse character, the lived world is closer to the oscillating realm of dreams than scientific descriptions.

In order to distinguish the lived space from physical and geometrical space, it can appropriately be called "existential space." Existential space is structured by meanings, intentions and values reflected upon it by an individual or a group, either consciously or unconsciously; existential space is a unique quality interpreted through human memory and experience. On the other hand, groups or even nations share certain characteristics of existential space that constitute their collective identities and sense of togetherness. The experiential and lived space, not physical or geometric space, is also the ultimate object and context of both the making and experiencing of architecture. The task of architecture is "to make visible how the world touches us," as Merleau-Ponty wrote of the paintings of Paul Cézanne.[19] In accordance with this seminal philosopher, we live in the

Figure 6.5 Design for a family focus. Cylindrical fireplace in a renovated attic flat, Helsinki, 1993. Blued steel. Photo Rauno Träskelin.

"flesh of the world" and architecture structures and articulates this very existential flesh, giving it specific meanings. I wish to suggest that architecture tames and domesticates the space and time of the flesh of the world for the purposes of human habitation. We know and remember who we are, and where we belong fundamentally through human constructions, both material and mental.

The World and the Mind: Boundaries of Self

"How would the painter or the poet express anything other than his encounter with the world?"[20] writes Maurice Merleau-Ponty. How could the architect do otherwise, we can ask with equal justification.

Art and architecture structure and articulate our being-in-the-world, or the inner space of the world (*Weltinnenraum*),[21] to use a beautiful notion of Rainer Maria Rilke. A work of art does not mediate conceptually structured knowledge of the objective state of the world, but it renders possible an intense experiential and existential self-knowledge. Without presenting any precise propositions concerning the world or its condition, art focuses our view on the boundary surface between our sense of self and the world.

In the text that he wrote in memory of Herbert Read in 1990, Salman Rushdie writes about the weakening of this boundary that takes place in an artistic experience: "Literature is made at the boundary between self and the world," he writes, "and during the creative act this borderline softens, turns penetrable and allows the world to flow into the artist and the artist flow into the world."[22] In fact, as we feel confident, protected and stimulated enough to settle in a space we allow similarly the boundary between ourselves and the space to soften and become sensitized.

All art articulates this very boundary surface both in the experience of the artist and the viewer. This is the zone of mental existential comfort. In this sense, architecture is not only a shelter for the body, but it is also the contour of the consciousness, and an externalization of the mind. Architecture, or the entire world constructed by man with its cities, tools and objects, has its mental ground and counterpart. The geometries and hierarchies expressed by the built environment, as well as the value choices that they reflect, are always mental structures before their materialization in the physical environment. Our most commonplace acts give evidence of inner mental landscapes, as inevitably as the rituals and monuments that we hold in highest esteem. Precisely, our most commonplace acts, to which we place least amount of conscious attention and embellishment, provide most conclusive evidence of our mental landscape. A setting wounded by thoughtless acts of man, fragmentation of the cityscape, as well as insensible buildings, are all external monuments of an alienation and shattering of the human inner space. "In accordance with the Almighty, we make everything in our own image, because we do not have a more reliable model; the objects produced by us describe us better than any confession of faith," writes Joseph Brodsky in his remarkable book *Watermark*, that analyzes touchingly the writer's experiences of Venice.[23]

Embodied Consciousness

Human consciousness is an embodied consciousness, the world is structured around a sensory and corporeal center. "I am my body,"[24] Gabriel Marcel claims, "I am what is around me,"[25] argues Wallace Stevens, and "I am the space, where I am,"[26] establishes the poet Noel Arnaud. Finally, "I am my world,"[27] concludes Ludwig Wittgenstein.

The senses are not merely passive receptors of stimuli, and the body is not a mere point of viewing the world through a central perspective.

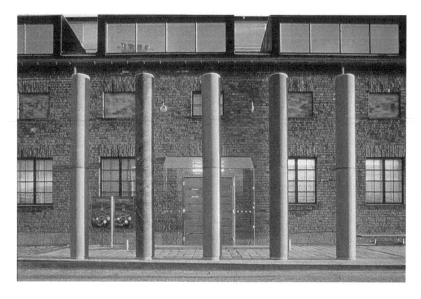

Figure 6.6 An architectural gesture. Colonnade in front of the main entrance, Rovaniemi Art Museum, renovation of a postbus depot to a museum of modern art, 1986. Source: Author.

Figure 6.7 Architecture and landscape. An artist's summer atelier, Vänö Islands, 1970. Photo Juhani Pallasmaa.

Our entire being in the world is a sensuous and embodied mode of being. The senses and our bodily being structure, produce and store silent knowledge. All our senses as well as our very being think.

In fact, the knowledge of traditional societies is stored directly in the senses and muscles instead of being a knowledge molded into words and concepts. Learning a skill is not founded on verbal teaching but rather on the transference of skill from the muscles of the master to the muscles of the apprentice through sensory perception and mimesis. The same principle of embodying – or introjecting, to use a notion of psychoanalysis – knowledge and skill continues to be the core of artistic learning. The foremost skill of the architect is, likewise, turning the multi-dimensional essence of the design task into an embodied image; the entire personality and body of the architect becomes the site of the problem. Indeed, architectural problems are far too complex and existential to be dealt with solely in a conceptualized and rational manner.

Art and Emotion

Architecture as with all art mediates and evokes existential feelings and sensations. Buildings of our time, however, have normalized emotions and usually completely eliminate such extremes of the scale of emotions as sorrow and bliss, melancholy and ecstasy. On the other hand, the buildings of Michelangelo, for instance, represent a touching architecture of melancholy and sorrow. But his buildings are not symbols of melancholy; they actually mourn. These are buildings that have fallen in melancholy – or more precisely – encouraged by their authority – we lend these buildings our own sensation of metaphysical melancholia.

In the same way, the buildings of Louis Kahn are not metaphysical symbols; they are a form of metaphysical meditation through the medium of architecture that leads us to recognize boundaries of our own existence and to deliberate on the essence of being. They direct us to experience our own existence with a unique intensity. Similarly, the masterpieces of early modernity do not merely represent optimism and love of life through architectural symbolization; even decades after these buildings were conceived, they evoke and maintain these positive sensations; they awake and bring forth the hope sprouting in us. Alvar Aalto's Paimio Sanatorium of the early modern era is not only a metaphor of healing; even today it offers the comforting promise of a better and more humane future.

True architectural quality does not derive from a formal or aesthetic game or from physiological and ergonomic parameters; it arises from experiences of an authentic sense of life, and an architectural structure can move us only if it is capable of touching something buried deeply in our embodied memories. This argument makes today's obsessive interest in innovation and novelty questionable. The temporal sense of architecture is engaged with our past and the human historicity as significantly as it is concerned with the future.

The Task of Art

As the consumer and media culture of today consists of increasing manipulation of the human mind in the form of thematized environments, commercial conditioning and benumbing entertainment, art has the ethical mission to defend the autonomy of individual experience and provide the existential ground for the human condition. One of the tasks of art and architecture is to safeguard the authenticity of the human experience.

The settings of our lives are irresistibly turning into a mass-produced and universally marketed kitsch. In my view, it would be ungrounded idealism to believe that the course of our obsessively materialist culture could be altered within the visible future. But it is exactly because of this critical view that the ethical task of artists and architects, the defense of the authenticity of life and experience, is so important. In a world where everything is becoming similar and, eventually, insignificant and of no consequence, art has to maintain differences of meaning, and in particular, the true criteria of experiential quality.

"My confidence in the future of literature consists in the knowledge that there are things that only literature can give us, by means specific to it,"[28] writes Italo Calvino in his *Six Memos for the Next Millennium*, and continues (in another chapter):

> In an age when other fantastically speedy, widespread media are triumphing, and running the risk of flattening all communication onto a single, homogenous surface, the function of literature is communicating between things that are different simply because they are different, not blunting but even sharpening the differences between them, following the true bent of written language.[29]

In my view, the task of architecture is to maintain the differentiation and qualitative articulation of existential space. Instead of participating in the process of homogenization of space and further speeding up of human experience, architecture needs to slow down experience, halt time and defend the natural slowness of human perception. Architecture must defend us against excessive stimuli, noise and speed. Yet, the most profound task of architecture is to maintain and defend silence. "Nothing has changed the nature of man so much as the loss of silence," warns Max Pickard, the philosopher of silence.[30]

"Only if poets and writers set themselves tasks that no one else dares imagine will literature continue to have a function", Calvino states. "The grand challenge for literature is to be capable of weaving together the various branches of knowledge, the various 'codes' into a manyfold and multifaceted vision of the world."[31]

Confidence in the future of architecture can, in my view, be based on the very same knowledge; existential meanings of inhabiting space can be wrought by the art of architecture alone. Architecture continues to have a great human task in mediating between the world and ourselves providing a horizon of understanding our existential condition and constructing settings for existential comfort.

Notes

1. Gaston Bachelard, *The Poetics of Space*. Boston: Beacon Press, 1969, p. 46.
2. Jean-Paul Sartre, *The Emotions: An Outline of a Theory*. New York: Carol Publishing Co., 1993, p. 9.
3. Jorge Luis Borges, *Selected Poems 1923–1967*. London: Penguin Books, 1985. As quoted in Sören Thurell, *The Shadow of A Thought – The Janus Concept in Architecture*. Stockholm: School of Architecture, The Royal Institute of Technology, 1989, p. 2.
4. Italo Calvino, *Six Memos for the Next Millennium*. New York: Vintage Books, 1988, p. 57.
5. Maurice Merleau-Ponty, "The Film and the New Psychology," in Maurice Merleau-Ponty, *Sense and Non-Sense*. Evanston: Northwestern University Press, 1964, p. 48.
6. Edward Relph, *Place and Placelessness*. London: Pion Limited, 1976, p. 51. Relph defines the notion as follows: "Existential outsideness involves a self-conscious and reflective uninvolvement, an alienation from people and places, homelessness, a sense of the unreality of the world, and of not belonging."
7. Joseph Brodsky, *Less than One*. New York: Farrar Straus Giroux, 1986, p. 124.
8. As quoted in David Michael Levin, editor, *Modernity and the Hegemony of Vision*. Berkeley: University of California Press, 1993, p. 14.
9. Ashley Montagu, *Touching: The Human Significance of the Skin*. New York: Harper & Row, 1968 (1971), p. 3.
10. Kent C. Bloomer and Charles Moore, *Body, Memory and Architecture*. New Haven, CT: Yale University Press, 1977, p. 44.
11. Marcel Proust, *In Search of Lost Time, Volume 1: Swann's Way*, translated by C.K. Scott Moncrieff and Terence Kilmartin. London: Vintage, 1996, pp. 4–5.
12. Gaston Bachelard, *The Poetics of Space*. Boston: Beacon Press, 1969, p. XXXIV.
13. Maurice Merleau-Ponty, "Cézannes's Doubt," in Merleau-Ponty, *Sense and Non-Sense*. Evanston: Northwestern University Press, 1991, p. 15.
14. Bernard Berenson, as quoted in Ashley Montagu, *Touching: The Human Significance of the Skin*. New York: Harper & Row, 1986, pp. 308–309.

    Somewhat surprisingly, in my view, Merleau-Ponty *objects strongly* Berenson's view: "Berenson spoke of an evocation of tactile values, he could hardly have been more mistaken: painting evokes nothing, least of all the tactile. What it does is much different, almost the inverse; thanks to it we do not need a 'muscular sense' in order to possess the voluminosity of the world [. . .]. The eye lives in this texture as a man lives in his house." Maurice Merleau-Ponty, "Eye and Mind," *The Primacy of Perception*. Evanston: Northwestern University Press, 1964, p. 166.

    I cannot, however, support this argument of the philosopher. Experiencing the temperature and moisture of air and hearing the noises of carefree daily life in the erotically sensuous paintings of Matisse or Bonnard one is confirmed of the reality of ideated sensations.
15. As quoted in Scott Poole, "Pumping Up: Digital Steroids and the Design Studio," unpublished manuscript, 2005.
16. See *Walter Benjamin's Philosophy: Destruction and Experience*, edited by Andrew Benjamin and Peter Osborne. London: Routledge, 1994.

17. Mildred Reed Hall and Edward T. Hall, *The Fourth Dimension in Architecture: The Impact of Building Behaviour.* Santa Fe: Sunstone Press, 1995.

18. Martin Heidegger, "Building Dwelling Thinking," in David Farrell Krell, *Martin Heidegger: Basic Writings.* New York: Harper & Row, 1997, p. 334.

19. Maurice Merleau-Ponty, "Cézanne's Doubt," op. cit., p. 19.

20. Maurice Merleau-Ponty, *Signs,* as quoted in Richard Kearney, "Maurice Merleau-Ponty," *Modern Movements in European Philosophy.* Manchester: Manchester University Press, 1994, p. 82.

21. Liisa Enwald, "Lukijalle," *Rainer Maria Rilke, hiljainen taiteen sisin: kirjeitä vuosilta 1900–1926* [The silent innermost core of art: letters 1900–1926]. Helsinki: TAI-teos, 1997, p. 8.

22. Salman Rushdie, "Eikö mikään ole pyhää?" [Isn't anything sacred?]. Helsinki: Parnasso, 1996, p. 8.

23. Joseph Brodsky, *Watermark.* London: Penguin Books, 1992, p. 61.

24. As quoted in "Translator's Introduction" by Hubert L. Dreyfus and Patricia Allen Dreyfus in Merleau-Ponty, *Sense and Non-Sense.* Evanston: Northwestern University Press, 1964, p. XII.

25. Wallace Stevens, "Theory," in *The Collected Poems.* New York: Vintage Books, 1990, p. 86.

26. In Gaston Bachelard, *The Poetics of Space.* Boston: Beacon Press, 1969, p. 137.

27. Ludwig Wittgenstein, source unidentified.

28. Italo Calvino, *Six Memos for the Next Millennium,* op. cit., p. 1.

29. Ibid., p. 112.

30. Max Picard, *The World of Silence.* Washington, DC: Regnery Gateway, 1988, p. 221.

31. Calvino, op. cit., p. 45.

Chapter 7

Outdoor Environment Indoor Space

Boon Lay Ong

Taking an international flight is much more disruptive to our bodies than most of us are aware of (McIntosh *et al.* 1998; Hinninghofen and Enck 2006; Spengler and Wilson 2003). Currently, over two billion people, equivalent in number to one-third of the earth's population, take a commercial flight in a year. We are familiar with minor problems like swelling joints and feet, stuffed ears and aching muscles. Dehydration is also a common problem leading not just to dryness in our skin but also to dry and irritated eyes. We arrive after a long flight exhausted rather than refreshed, despite adopting a position that might be considered to be restful and being entertained, fed and catered to. Our body clock does not easily adapt to a different time zone and the larger the difference in time zones the greater the stress on the body. Often, arriving in a different country means coping with different cultures and expectations as well. The local language can be different and so there is a linguistic adaptation on top of the cultural and biological changes. It is as great and as sudden a cultural and physical shock as we are ever likely to experience.

Arriving at the airport after a flight is therefore an important transitional episode in the phenomenology of flying. Interestingly, while physically in a different country, the actual act of entering the country only happens when you pass through the immigration gantry. Many of us take this opportunity to relieve ourselves and freshen up before facing the immigration officer, collecting our luggage, passing through customs and then meeting up with friends or relatives, or else catching a cab or some other form of transport to a hotel.

Most international airports are nationally important buildings designed with considerable thought and expense focusing on making an impression, imparting a sense of national identity and efficiency of use. Despite this stature, the sense of welcome and comfort at international airports is one of garish ostentation rather than the repose and warmth that the traveller most needs. The facility most visited by incoming travellers and which can most easily and lastingly leave a bad impression is one that is least carefully designed and maintained – the restroom. The restrooms in most, if not all, international airports are entirely enclosed, windowless, sterile and harsh environments poorly designed for refreshing ourselves. They are, quite simply, just toilets.

An opportunity arose to reconsider the design of airport restrooms in the tropics. Although not intended for actual construction, the exercise enabled us[1] to rethink the process of flying and arriving at a new country and the environmental conditions that should prevail. After a round of brainstorming, we decided that a good approach and challenge will be to do the opposite of what the existing facility was designed to be – give the user a sense of what the environment that lies outside and awaits him is like.

The Phenomenology of Arrival

Touchdown is a tense and anxious time for the passengers. For many, the flight has been a very long separation from home – physically, mentally and emotionally. This separation has been filled with a further sense of suspension as one is neither in one country nor the next. Flying, very literally, displaces us. WHO (2010) identifies arrival not only with culture shock but also with a reverse culture shock when one returns from an extended stay in a foreign country. There is a sudden bustle in the cabin when passengers are told that they may leave the aeroplane. Until that moment, we are not just in our seats but strapped down. Some make a conscious effort to be patient but nearly everyone is eager to be out of the plane. In long flights, the windows are drawn and the only clues we have about time are our own body clocks and cabin lighting. If we watch inflight movies, we are even further removed from the ebb and flow of day and night. If we can sleep in the airplane, we prefer night flights so that we arrive in the morning a little rested and our body clocks somewhat readjusted. Even under the best of conditions, however, experts (Duffy and Czeisler 2009; Arendt 2009) recommend allowing at least one day for each hour of change in time zone and preparing for this change several days pre-flight. It is a luxury that not many can afford.

The air quality in the cabin is more unhealthy than commonly realised (Spengler and Wilson 2003). The low air pressure causes nitrogen to expand in our blood and our joints to bloat. The low humidity (lower than deserts on the ground) causes rapid dehydration and our bodies to readjust its internal biochemical levels. The carbon dioxide level is three to four times higher than on the ground. Cramped together inside the womb of an aircraft, we surge out into the airport eager to reconnect with land and society. All international airports today proclaim their modernity and technological advancement but feel more like commercial departmental stores than communal places. For visitors, the character of a country is to be found in the public social places. Very few visitors enjoy the privilege of staying in a local home (bed and breakfast lodgings notwithstanding) and most of us imbibe the local flavour through the streets, the eating places and the shops. Arriving at the airport is a hasty affair, we are glad to leave the stuffiness of the plane and eager to be through the gates into the 'real world' beyond. Our minds are already on the other side of the gates, thinking about the people we are meeting, wondering what the place we are visiting might be like (or if we are returning, wondering what has changed since), beginning to adjust to the different climate that we are in. Not many linger, except to buy a gift from the duty-free shop or to head for the restroom for some badly needed respite. As soon as allowable, mobile phones are activated to make at least a reassuring vocal contact with the outside world.

Stress and Place

Stress is an issue that is not covered elsewhere in this book but is one that intimately impacts all the discourse within. First discovered in the scientific world by Hans Selye in the 1970s (Cooper and Dewe 2004), it has since become the main protagonist in urban and modern health and well-being. It is difficult enough to go beyond environmental comfort and touch upon realms of culture, emotion, psychology and aesthetics without tackling the dynamic nature of human behaviour. The concept of comfort suggests repose and rest and in defining environmental criteria around this concept, we tend to avoid the inconvenient fact that our bodily needs are temporal, changing with the time of day, seasons, events and personal circumstances. Acknowledging that our individual needs change over time requires us to further acknowledge that the environ-mental needs of each of us differ one from another. With such a diversity of needs, it becomes almost impossible to define environmental criteria to any useful purpose. Fortunately, the reality is not as dire as we might imagine.

What the essays in this book point to is that our different needs are held together in the phenomenon of place. As much as our immediate needs may be affected by our past activities, our physical bodies and climate, the components of place also guide us towards a common 'comfort' zone appropriate to the place. We behave differently in different places because there are environmental cues that prompt us towards the appropriate level of stimulation. These cues are particularly important in the phenomenology of flying. If excited, the right environment can calm us down. If tired, the right environment can energise us. Until now, we were not aware that any airport design had taken these considerations on board. This was an opportunity for us to see how such considerations can impact design, particularly in terms of helping travellers synchronise their biological clocks.

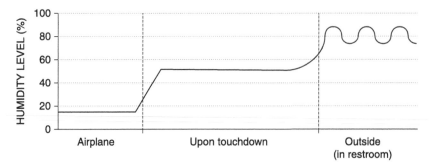

Figure 7.1 Changes in humidity levels as an indication of climatic adaptation from an airplane to the tropical conditions outside (and the proposed restroom).

Climatic and Biological Adaptation

In designing the internal environment of an airport, very little thought is given to providing an interface between the plane and the environment outside the airport. Yet, adaptation to the time and weather of a new location is perhaps the hardest adaptation we have to make. Our exhausted bodies respond slowly to the new clime and the visual and temperature cues in the airport are not helpful in this regard. Due to the controlled environments of both the airplane and the airport, there is very little in common between the airplane environment, the airport environment and the outside environment. Humidity levels are especially important (Figure 7.1) and cause most of the noticeable effects of airplane travel.

The airport restroom is an important part of this adaptation. Even under current conditions, a visit to the restroom can be refreshing through simple face washing and grooming. A key concern in our design approach is to provide a sense of the outside environment inside the restrooms so that a stronger sense of arrival is felt and the visitor better prepared for the other side of the airport gates.

Conceptual Approach

From the very start, we decided that the most important component in the design is the inclusion of greenery. Plants link us to place in very many ways. Natural vegetation is specific to the location and we recognise them quickly, for most of us unconsciously and without being able to name actual species nor explain why and how (Riegner 1993). We instinctively recognise greenery to signify a habitable environment and to identify specific climatic conditions. Research has found that greenery is not only calming but critical in cognitive and attentive development in children as well as in corresponding performances in adults (Kaplan 1995; Wells 2000; Taylor *et al.* 2001). Seamon (2000) argues that phenomenologically, people and the environment are an indivisible whole. If so, this phenomenological wholeness is ruptured by flying and needs to be reinstated as soon as possible upon arrival. This wholeness with the environment is closely linked with our biophilic relationship with plants. Jack Forbes (2001) puts it bluntly:

I can lose my hands and still live. I can lose my legs and still live. I can lose my eyes and still live . . . But if I lose the air I die. If I lose the sun I die. If I lose the earth I die. If I lose the water I die. If I lose the plants and animals I die. All of these things are more a part of me, more essential to my every breath, than is my so-called body. What is my real body?

There are strong ecological themes in our wholeness with plants. Plants are our ultimate source of food, provide shade and shelter, cleanse the air and water and provide material resources for medicine, construction and clothing. We are dependent upon plants so much that they infiltrate our emotional and mental well-being. Plants are themselves ecological and present us with physical references to time, season and place. Adopting plants as our central motif in the design inescapably engages us with an ecological approach in both concept and detail development.

Finally, plants are the ultimate panacea for stress (Kaplan 1995; Ulrich 2002; Kellert *et al.* 2011). By incorporating plants as a main component of our design, we engage with these issues of place, biorhythm and stress without having to explicitly address these complex issues. Research into the benefits of plants is well established and widely acknowledged. Indeed, it is scientifically unenlightened today to design a healthcare facility without providing access to gardens and views to greenery. The intimacy and comprehensiveness by which plants provide for our well-being and healing are difficult to quantify and isolate as researchable attributes. Most of the evidence accumulated points to the general benefit of plants but cannot go much beyond that to suggest what kinds of plants are more beneficial, for what purposes and under what conditions. Commonsensical strategies understood through our history of gardening and socio-cultural traditions about plants help us more than science can beyond urging for the widespread adoption of greenery.

Biologicalrhythms

Environmental comfort in today's literature is almost entirely focused on static and constant criteria. For example, the recommended lighting level for reading may be 500 lux, depending on which standard you refer to (see Chapter 1), but this lighting level does not take into account the time of day. In the same way, recommended noise levels and temperatures are not aligned to the diurnal rhythm, let alone seasons. And yet, we know ourselves that our body ebbs and flows over the course of the day, our attention span varies, our resistance to both physical viruses and mental suggestions change, and our emotions and moods too change with both time and the environment. Activities in the morning affect our performance in the afternoon and our mood at night. Where we were or what we were doing an hour ago affects what we need for environmental comfort now. The dynamic nature of human comfort is woefully neglected in environmental research.

Air travel brings the problems with biorhythms to the fore. Whilst in everyday living, we take circadian cycles and biological adaptations in our stride, the strain of changing time zones is more than most of us can tolerate. This project offered us an opportunity to engage with biological rhythms as a design driver and in many ways open up new areas for future exploration in environmental research.

Our premise became our central problem: how to create an environment conducive to plants and people in a totally serviced space. Plants as living things require food, water, light and air. Plants also respond to circadian as well as annual cycles. Seasons in the tropics are not as markedly variable as in temperate countries and changes just from dry to wet. In an equatorial environment like Singapore (Figure 7.2 and Figure 7.3), the

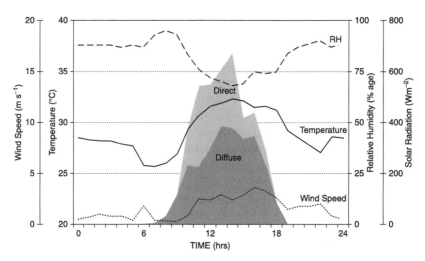

Figure 7.2 Typical daily weather data for Singapore (air temperature, relative humidity, direct and diffuse solar radiation and wind speed). Source: Author.

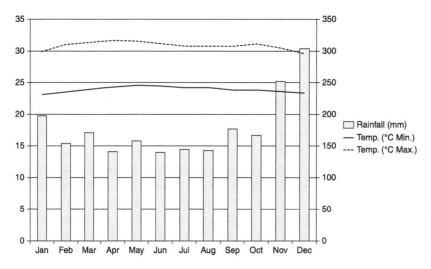

Figure 7.3 Singapore annual climate data (minimum and maximum temperature, monthly rainfall). Source: www.metoffice.gov.uk.

variation in rainfall and humidity throughout the year is even less than else-where in the tropics. Over the span of a day, air temperatures range from around 25°C during the night to 35°C in the afternoon while relative humidity is around 85% at the lowest and above 95% at the highest. Solar radiation is generally diffuse with above 70% cloud cover most of the time. Diffuse radiation accounts for around one third of the total solar radiation received.

The stable climate of Singapore means that the greatest concern in terms of mimicking biological rhythms is the circadian cycle. The circa-dian cycle for humans is actually similar to that for plants and other animals. However, as a result of artificial lighting as well as the popularity of inter-national flights, we live our lives less in synchronicity with the natural cycle of night and day than we should. Circadian cycles affect us more than just our alertness and sleeping hours. The severity of many common illnesses varies with time of day (Litinski *et al.* 2009: 143):

> . . . myocardial infarction occurs most frequently in the morning a few hours after waking up, epileptic seizures of the brain's temporal lobe usually occur in the late afternoon or early evening, and asthma is generally worse at night. There are also differ-ences across the 24-hour period in cancer development and on chemotherapeutic effectiveness. In addition, shift work is generally associated with chronic misalignment between the endogenous circadian timing system and the behavioral cycles, including sleep/wake and fasting/feeding cycles, and this misalignment could be a cause of the increased risk of diabetes, obesity, cardiovascular disease, and certain cancers in shift workers.

Designing for Plants

It is difficult to ensure that plants will grow well in any environment, much less inside an airport restroom. Despite this vulnerability, it is necessary that the plants in the restroom are well maintained and aesthetically pleasing at all times. In order to ensure this, we developed a module for the plants that will allow them to be removed and replaced as needed. There were two types of modules – one, a slim vertical wall and two, a deeper tray module. The slim version, *living screen*, is 200 millimetres deep and can act as a wall by itself. The deeper version, *living cell*, is 600 millimetres deep, allowing the inclusion of services and meant to be inte-grated with the water tanks of toilet units. The living screen can be config-ured in different ways to allow for deeper soil, water base and even vertical soil media. If necessary, media display modules can also be installed instead of plants.

The soil media used is lightweight aggregate in most instances. This will help reduce insect infiltration and ensure a generally sterile

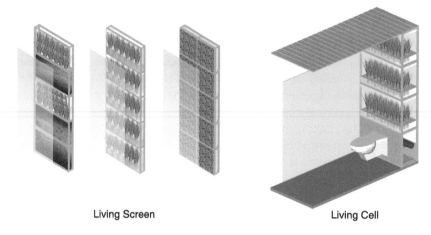

Living Screen   Living Cell

Figure 7.4 Two modular systems proposed for the airport toilet design.

environment needed in an airport setting. In vertical planting, the base wall is lightweight aggregate held together by coir or jute geotextiles. Moss or lawn grass is proposed so that a more refined and carpet-like finish is achieved. Where the plants can be grown to a horizontal bed, herbs like mint, lemongrass or oregano may be used to provide a gentle perfume to the air inside the restroom.

The plants are grown either in small removable pots or vertical 900 × 900 millimetre modules so that they can be replaced as often as required. The enclosures for the plants, both of the screens and the cells, are porous to allow the restroom air to be refreshed by the oxygenated and fragranced air of the plants. Lighting for the plants is a mixture of fluorescent and metal halide lamps to provide the right temperature and colour for healthy plant growth. The plant enclosures are misted at 15-minute intervals to maintain a higher humidity within the enclosures.

The Lighting Regimen

The environment in the restroom has to simulate the external night–day cycles to maintain the health of the plants. As mentioned earlier, maintaining the circadian cycle is important for human beings as well. At the same time, there needs to be enough light for the users to see their way around and find the facilities at night.

Several strategies were used. First, different lighting rhythms were programmed for the plants and for humans. The lighting levels for the plants will fluctuate over the day to rise to a maximum during the afternoon to follow the daylighting levels outside. At night, the plants will prefer a dark environment and so only accent lighting will be used to showcase the plants through colour rather than brightness. The lighting strategy for humans will also fluctuate over the day but less so – less bright during the daytime but brighter than in the plant enclosures at night. In the daytime, the ideal proportion is for the restroom to feel indoors while the plant enclosures need to be bright enough to feel like outdoors. During the night, the ambient lighting will be kept low enough to give the user a sense of night-time. Dawn and dusk lighting are lower and tinted with shades of orange, violet, red and yellow as the low angles of the morning and evening sun's light are refracted by the clouds and atmosphere. At night, the living screens and cells need just enough lighting to be seen as a display. This is achieved using vertical lights just outside the screens and keeping the levels within the plant enclosure at night-lighting levels (Chaney 2005). Both low levels as well as colour composition of the lighting are important. In order of preference, Chaney recommends mercury vapour, metal halide or fluorescent. At night, attention is drawn more to the facilities themselves, which can also be attractive in their own way. One strategy to provide light only where it is needed is to create a runway atmosphere by running recessed lights on the floor of the restroom to help in wayfinding as well. Compensatory light, if needed, is provided particularly in toilet cubicles during the night (Figure 7.5).

In all, there are four different lighting strategies: lighting for the plants, lighting for human circadian cycles, functional lighting (including wayfinding) and lighting for accentuation of plants and other displays.

Humidity Control

Plants require a higher humidity than normally found in airports to maintain turgidity. Human skin as well fares better at higher humidities, provided that the air is cool and sweating is not a problem. The humidity in the restroom is designed to fluctuate between 70 to 85 per cent RH with regular misting. The misters are located within the plant enclosures and programmed to mist every 15 minutes. Ultrasonic humidifiers are used to

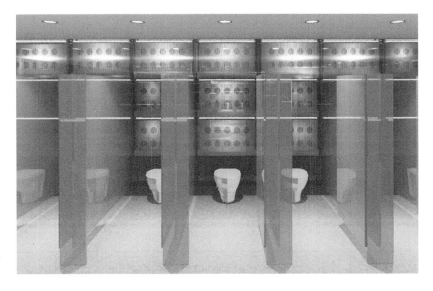

Figure 7.5 Lighting strategy for airport restroom.

provide a very fine mist that will evaporate quickly and not leave wet condensation on the skin.

Air-Conditioning and Ventilation

The restroom is completely air conditioned. Fresh air is provided through mixture in the air-conditioning system and no opening to the outside is available for natural ventilation. One of the important purposes of air conditioning is to remove the smells associated with restrooms. Most restrooms are ventilated at the ceiling because it is more convenient and cost effective to locate the air-con ducts there. Smells in the restroom are usually masked by regular bursts of fragrance from perfume dispensers. In the proposed restroom, the flow of air in the restroom is designed to come in at the top and removed at floor level. Additionally, the inlet air is channelled through the plants to take up additional oxygen produced by the plants during the daytime. The plant enclosures have large circular openings to allow passengers to rub the plants and perhaps pluck a small leaf to release the plant's fragrance. Despoiling the plants through excessive plucking can be simply offset by replacement.

Design Development

While desiring to provide a natural environment indicative of the world outside, it was felt to be necessary also to design the restroom to reflect the modernity and technological advancement of the airport itself. This was achieved through the use of stainless-steel detailing and glass. The architectural language was minimalistic and colour was primarily reserved for the plants.

The layout of the restroom is more luxurious than normally provided as passengers are expected to spend more time in it than in other restrooms. The function of the restroom is more than just toilet but a more

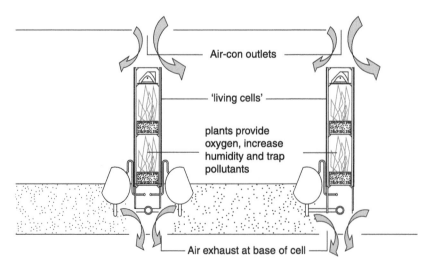

Figure 7.6 Ventilation flow of fresh air through the plant enclosures to provide fresher air and maximise the benefits of the plants.

Figure 7.7 Waiting and refreshment area within the restroom.

comprehensive refreshing of oneself. A waiting area, in particular, was felt to be necessary so that queuing can be avoided. Seats are provided and passengers, particularly women, can touch up their makeup or comb their hair while waiting. A polyurethane floor finish with a light pebble pattern helps to give the restroom a more natural feel while keeping maintenance to a minimum. Walls are mirrored to increase the spaciousness of the restroom and to further facilitate grooming.

Implications of Biorhythm Design

The success of this design, however, requires a supportive doctrine to be applied to the rest of the airport. The thought processes engaged with here suggest that the whole flight experience can be, and should be, designed to ease the passenger through the transition between time zones. Would not an ambience more reflective of the time and season help put the arriving passenger into the right frame of mind? The bustle of an airport in the morning and daytime is usually sufficient to wake the arriving passenger and gear him up for an active day. But can we not similarly provide soothing and quiet transitions so that the biorhythms of the passenger are subtly calmed and soothed so that he is more easily lulled to sleep when he finally reaches his destination? Inasmuch as adaptive thermal comfort takes into account the climate in different places, is it so difficult to extend this idea to take into account different times of day and also to apply to different areas of environmental science – adaptive lighting, adaptive noise control, etc.? It is not difficult. It simply requires us to design to aesthetic or adaptive standards rather than technical ones.

Airports, then, are not destinations in themselves. They are vital transition places that need to be designed to best provide the environmental cues and conditions to help the visitor adapt to local conditions. At the very least, arriving and processing in the airport can take half an hour. Often longer. This is not an insignificant stretch of time. Much can be done during this time to help the body transition to the local time zone. Adopting an adaptive approach, which more closely follows external conditions, will result in less energy consumption and waste as well. It is less stressful to the body and can help ameliorate the stress and health problems of long-distance flights. Welcoming a guest does not always mean loud noises and bright lights. A warm welcome and a quiet embrace can be just as, if not more, welcoming.

Conclusion

Architectural aesthetics differs from art in that architecture is immersively experienced. As we move within and use a space, the aesthetic experience we savour is one that engages our bodies intimately and evokes physical responses that not only affect our comfort but also our well-being and health. Several chapters in this book reflect upon this aesthetic from different perspectives. This particular project expands yet further on the approach advocated by the various authors in this book. The views presented here are already on the cutting edge – realms of investigation considered outside the mainstream and where rigorous research is still in its infancy. Even at this cutting edge, the problem of dynamism or change in environmental comfort is not considered or discussed. The need for science to catch up with human needs in all its complexity cannot be overstated.

Aesthetics in architecture as well falls far behind from fully meeting the needs of the human occupant. Modern aesthetics is preoccupied with the visual and treats architecture as an object, often viewed from a distance. The aim of this exercise is to argue that an understanding of environmental aesthetics, going beyond environmental comfort, does more than provide good visual aesthetics. It goes to the heart of how we use and respond to the environment. The core understanding of heat, light and sound is just a foundation upon which much else needs to be built. Additional sources of knowledge and insights are taken from other fields like medicine, phenomenology, horticulture and human behaviour.

The design of the airport restroom began with a phenomenological accounting of what arrival at an airport after a long flight means to the traveller. This accounting, while phenomenological, draws upon more mainstream scientific research on circadian cycles and passenger health in airplanes as well. This inclusivity, however well done or not, is important and shows how architecture itself, over and beyond that provided by mechanical services, contributes to the health of its inhabitants. Increasingly, there is convincing evidence that this role of architecture of contributing to the health and well-being of its inhabitants requires the making of meaningful connections with the external environment. This in turn highlights the importance of the external environment to human life, even in an urban context or within an entirely artificial space.

A project as humble as an airport restroom can be significant if it is understood as part of the greater flow and movement between spaces and places in an airport. Indeed, for this project to be truly beneficial to the passengers, the circadian cycle of the restroom needs to be reflected in the design of the airport itself and ultimately integrated with the lighting and environmental programming of the flight itself. Arendt (2009) suggests that preparation for jet lag should allow for one day per hour of time-zone change and with lighting in the airplane itself timed to help our bodies make the time shift. Light bursts, as short as five minutes, can affect the synchronisation of our biological clocks. Further, it is not the actual light levels but the relative light levels that are important. The real challenge in architecture is to understand how knowledge gained through research in various fields can be translated into building design. This translation is not separate from the standards in lighting, heat and sound in current use, it is in addition to and integrated with standard practice. But there is also a reverse challenge. Good architecture functions well beyond the limits of what we currently understand through research. It is also by contemplating the gaps revealed when we consider the functioning of architecture that we can be directed towards new research and understanding.

Note

1. 'Us' is a team of lecturers from the School of Architecture, National University of Singapore, during the early 2000s. The team comprises Philip Bay Joo-Hwa, Alan Woo, Stephen Wittkopf and the author.

References

Arendt, J. (2009) Managing jet lag: Some of the problems and possible new solutions. *Sleep Medicine Reviews*, 13: 249–256.

Chaney, W. R. (2005) Does night lighting harm trees? *TurfGrass Trends*, September, 65–69.

Cooper, C. L. and Dewe, P. (2004) *Stress: a brief history*, Oxford: Blackwell.

Duffy, J. F. and Czeisler, C. A. (2009) Effect of light on human circadian physiology. *Sleep Medicine Clinic*, 4: 165–177.

Forbes, J. D. (2001) Indigenous Americans: Spirituality and ecos. *Daedalus*, 130(4): 283–300.

Hinninghofen, H. and Enck, P. (2006) Review: Passenger well-being in airplanes. *Autonomic Neuroscience: Basic and Clinical*, 129: 80–85.

Kaplan, S. (1995) The restorative benefits of nature: Toward an integrative framework. *Journal of Environmental Psychology*, 15: 169–182.

Kellert, S. R., Heerwagen, J. and Mador, M. (2011) *Biophilic design: the theory, science and practice of bringing buildings to life*, Hoboken, NJ: John Wiley & Sons.

Litinski, M., Scheer, F. A. J. L. and Shea, S. A. (2009) Influence of the circadian system on disease severity. *Sleep Medicine Clinic*, 4: 143–163.

McIntosh, Iain B., Swanson, V., Power, K. G., Raeside, F. and Dempster, C. (1998) Anxiety and health problems related to air travel. *Journal of Travel Medicine*, 5: 198–204.

Riegner, M. (1993) Toward a holistic understanding of place: reading a landscape through its flora and fauna. In *Dwelling, Seeing, and Designing: Toward a Phenomenological Ecology*, D. Seamon, ed. Albany, NY: SUNY Press, 181–215.

Seamon, D. (2000) A Way of seeing people and place: phenomenology in environment–behavior research. Published in *Theoretical Perspectives in Environment–Behavior Research*, S. Wapner, J. Demick, T. Yamamoto, and H Minami eds. New York: Plenum, 157–178.

Spengler, J. D. and Wilson, D. G. (2003) Air quality in aircraft. Proceedings of the Institution of Mechanical Engineers, Part E: *Journal of Process Mechanical Engineering*, 217(4): 323–335.

Taylor, A. F., Kuo, F. E. and Sullivan, W. C. (2001) Coping with ADD: The surprising connection to Green Play Settings. *Environment and Behavior*, 33: 54–77.

Ulrich, R. S. (2002) Health benefits of gardens in hospitals. *Plants for people, International Exhibition Floriade 2002*. www.greenplantsforgreenbuildings.org/attachments/contentmanagers/25/HealthSettingsUlrich.pdf (accessed 7 July 2012).

Wells, Nancy M. (2000) At Home with Nature: Effects of 'Greenness' on Children's Cognitive Functioning. *Environment and Behavior*, 32: 775–795.

WHO (World Health Organization) (2010) International Travel and Health. www.who.int/ith/ITH2010.pdf (accessed 24 March 2011).

Chapter 8

The Poetics of Environment

Dean Hawkes

Introduction: Background and Method

Over many years the focus of my architectural research has been in the field that is now commonly known as environmental design. In the early years I was involved in the development of computer simulation models for environmental analysis and then, leading quite directly from this, in the construction of comprehensive conceptual models of what we called the "environmental system" of buildings. That led to field studies that aimed to discover more about the place of the occupants of buildings in the processes of environmental management. As time passed the nomenclature of the work evolved; questions of *environmental design* became questions of *energy demand*, then of *energy consciousness*, or *bioclimatic architecture*, *passive solar design* and, now, the catch-all category of *sustainable design*. Another shift reflected my growing interest in other questions and, in particular, the relationship between the principally technical discourse of environmental studies and the broader concerns of the history and theory of architecture. Much of this work is condensed in *The Environmental Tradition*.[1] There a collection of essays was organised under the two headings of *Theory* and *Design*. In both categories connections were indicated between contemporary research and practice and theories and practices of the past. References to the works of Vitruvius, Palladio and Alberti are found alongside discussions of Le Corbusier, Louis Kahn, Piano and Rogers and Robert Venturi. The research that is presented in the present essay is a further step in linking the technologies of the architectural environment, *technics*, to its qualities, the *poetics of environment*.[2]

This project began with the proposition that the complex sensory experiences that we enjoy in great buildings are the product of acts of *imagination* on the part of architects, which bring together the interaction of light and air and sound with the form and materiality of architectural space. To explore this idea the work began with documentary and archival studies that helped to define the scope of the inquiry and to establish specific themes to be studied in an extensive programme of fieldwork. The essence of the matter, however, can only be addressed through first-hand experience of buildings themselves. The works of Steen Eiler Rasmussen and Juhani Pallasmaa,[3] in their different ways helped to define the approach. With the benefit of their insights, a tentative method was developed, themes were identified and a research programme was drafted. Consequently, hours were spent immersed in the ambience of many buildings in Europe and North America. The process involved mechanical recording through notes, sketches and photographs, but the principal tool was the actual experience of a building as time elapsed and qualities of light, air and sound, individually and in ever-shifting combinations, were directly experienced. The only valid instrument was the human senses. What follows presents some of the results of this research.

Carlo Scarpa and the Environment of the Veneto

Carlo Scarpa (1906–1978) spent the whole of his life in Venice, the city of his birth, or in the surrounding region of the Veneto. Scarpa's built works are, with a few exceptions, concentrated in this fertile landscape between the Adriatic and the Dolomites and in its fabulous cities, Venice itself and, to the west, Vicenza and Verona. The work is rooted in the conditions of culture, history, climate and building traditions of the Veneto; the landscape of Palladio.

Writing about Scarpa's architecture Francesco Dal Co has observed that "[t]he sensitivity Scarpa reveals in, for instance, his treatment of light and its handling of colour tones is the outcome of . . . his profound affinities with Venice."[4]

Museo Canoviano at Possagno

The influence of this regional tradition of environmental response is extensively revealed in Scarpa's mature projects. In 1955 he began work at the Museo Canoviano, at Possagno, where, in commemoration of the bi-centenary of Antonio Canova's birth, he added an extension to the existing nineteenth-century gallery.

This is a key building in Scarpa's *oeuvre* and demonstrates fundamental aspects of his environmental intentions. Speaking about the project in 1976, Scarpa expressed the nature of his approach when he said, "I really love daylight: I wish I could frame the blue of the sky!"[5] Sergio Los has written at length about the narrative function of light in Scarpa's approach to the display of Canova's sculptures, "bringing them to light."[6] In the same essay he goes on to suggest that "[i]t is precisely the light that shows the illuminated sculptures and 'translates' Canova, giving them a new interpretation and constituting – together with the organisation of space and construction – the typological content of the museum."

Scarpa's building takes its place in an ensemble consisting of Canova's former house and garden and the neoclassical basilican gallery by Francesco Lazzari, 1832–1836. Scarpa's contribution is, in effect, an extension to Lazzari's building, through which it is entered. The emphatic juxtaposition of Lazzari's neoclassical symmetry and conventional architectural language with Scarpa's free compositional system lies at the heart of the project. The formal differences between these distinct modes of architectural composition are reinforced by their fundamentally different environmental qualities.

Lazzari's basilica, whose long axis is orientated north–south, is lit through small rooflights set at the apex of the coffered barrel-vaulted ceiling, one in each of the three bays. These shed a generalised light within the space, although direct sunlight enters with dramatic effect when the sun is high in the summer months. The diurnal symmetry of the light is quite different from that of a basilican church, where the east–west orientation demanded by Christian orthodoxy creates a strong contrast between north and south aspects. Here the uniformity of the light emphasises the geometrical axiality of the space. Scarpa's light couldn't be more different. From the vestibule of the basilica the eye is led into a dazzling white volume in which sculptures are freely disposed in a complex field of light, some in silhouette, others brightly illuminated.

Individual works by Canova are disposed in both the square "high hall" and in the sequence of spaces that open from it, descending southwards, in step with the contours of the site. The light works in concord with the individual works of art, defining their locations relative to each other and to the space they inhabit. Los has observed:[7]

> Each statue has a very precise place, with respect to the overall space and to the light that pours in – at times with glaring violence, at other times softly and faintly – modelling the plasters on display, modifying them over the course of the day, with the changing seasons and the variations in weather.

The extreme contrast between the Lazzari and Scarpa spaces, set adjacent to each other and, therefore, in exactly the same field of ambient light, derives from the totally different conceptions of architecture and its method that they represent, the one conventional and formal, the other inventive, specific and intuitive.

Scarpa's invention of the trihedral corner windows, simultaneously window and rooflight, that illuminate the "high hall" is, environmentally and tectonically, the most remarkable element of the building. To the west they are tall and concave and to the east, where they are partly obstructed by the mass of the basilica, they are cubic and convex. Their configuration admits light from all orientations and, unlike a conventional window set within a wall, casts light across the walls themselves. This apparently simple device is the source of the magical quality of light that binds art and architecture into a complex unity. Los has proposed that the Modernist invention of the corner window, as Scarpa adapted it, lies in lineal succession of the principles outlined by Palladio as the basis of *his* Veneto architecture of the sixteenth century.[8] The ratio of solid to void in the "high hall" at Possagno is as constrained and disciplined as Palladio's dimensioning of openings in the walls of both villas and palazzos.

Looking west into the "high hall" the afternoon light of an early summer's day is cast directly on the floor, the northern wall and the sculptures. In addition, it is diffracted onto the other wall surfaces by the trihedral windows. From a west-facing viewpoint the figure of George Washington, in senatorial costume, is silhouetted against the light, but is fully illuminated and modelled when viewed from the other side (Figure 8.1).

This image demonstrates the variety of conditions under which the sculptures are simultaneously illuminated. George Washington is

Figure 8.1 Carlo Scarpa, Museo Canoviano, Possagno. Interior.

strongly modelled by direct sunlight, "Amor and Psyche with Butterfly" stand in relative shade, against the projection of the light of the corner window upon the wall, and the bust of Napoleon is softly modelled against the shadow by reflections transmitted from the adjacent sunlit wall.

The famous group, the "Three Graces," is positioned at the end of the southerly extension of the gallery, in front of a mullioned window that rises above the soffit of the gallery. Outside a small pool reflects light onto the ceiling. Looking towards the window the figures are in silhouette, but from the reverse they are, in the afternoon, softly illuminated by diffuse light (Figure 8.2).

In his 1976 lecture Scarpa spoke of the issues behind the conception of this arrangement:

> I wanted to give a setting to Canova's "The Graces" and thought of a very high wall: I set it inwards because I wanted to get the light effect of a bay. That sort of dihedron getting into the room produces that fineness of light which makes that paint as well-lit as the other walls.[9]

It is clear from Scarpa's descriptions, that light was the primary element of the environmental vision of the project. Sculptures, by their nature, are environmentally robust and are not subject to the strict demands for conservation that apply to the design of spaces for the display of paintings and drawings. They are similarly unaffected, in relative terms, by the thermal environment in which they are kept.

It is rarely observed that the Gipsoteca has no heating system. It simply shelters its priceless contents from the elements by the methods adopted by all pre-industrial architecture, by the organisation of form and material, of solid and void. In the true sense of the term this architecture is *primitive*, "relating to . . . the character of an early stage in the evolutionary or historical development of something." But this only enhances the profound quality of the building by maintaining the essence of the issue. By rejecting the implicit technological expectations of mid-twentieth-century building, with its extensive apparatus of environmental services, Scarpa is able to focus on the fundamentals of the architectural environment. The ambience of the interior, luminous, thermal and acoustic, is the direct product of the interposition of the enclosure within the ambient climate. In this way, whatever the consequences in terms of modern notions of environmental comfort, there is an absolute unity of the internal environment in all its dimensions. This building is a supreme instance of the poetics of environment.

Sigurd Lewerentz: Architecture of Adaptive Light

One of the most remarkable attributes of human vision is our ability to see in levels of light that vary from the darkness of a moonless night to the

Figure 8.2 Carlo Scarpa, Museo Canoviano. *The Three Graces.*

vivid glare of a sunlit summer's day. This phenomenon is known as adaptation and the mechanisms by which it occurs and its implications for building design have been the subject of much study.

The psychologist Richard Gregory suggests:

> It might be said that moving from the centre of the human retina (cones) to its periphery (rods) we travel back in evolutionary time, from the most highly organised structure to a primitive eye which does no more than detect simple movements of shadows.[10]

A further, and specifically architectural, insight into the poetics of adaptation is offered by Juhani Pallasmaa in *The Eyes of the Skin*,[11] where he writes:

> The eye is the organ of separation and distance, whereas touch is the sense of nearness, intimacy and affection. The eye controls and investigates, whereas touch approaches and caresses. During overpowering emotional states, we tend to close off the distancing sense of vision; we close our eyes when caressing our beloved ones. Deep shadows and darkness are essential because they dim the sharpness of vision, make depth and distance ambiguous and invite unconscious peripheral vision and tactile fantasy.

Sigurd Lewerentz made designs for sacred buildings throughout his long life[12]; the two masterpieces of St. Mark's, Björkhagen, Stockholm, 1955–1964 and St. Peter's, Klippan, 1962–1966 were important subjects of these investigations.

St. Mark's Church, Björkhagen, Stockholm

In 1955 Lewerentz won the competition to build St. Mark's Parish Church at Björkhagen in the southern suburbs of Stockholm. The site is a small birch wood lying low in relation to the surrounding terrain. Lewerentz's design consists of two buildings, a low wing containing offices connected to the belfry and an L-shaped block in which a series of rooms for parish activities are linked to the church itself. The offices and parish rooms stand opposite each other defining a courtyard.

St. Mark's was the first building in which Lewerentz used brick as the predominant material. The interior of the church has walls of fair-faced brickwork and is covered with a series of brick vaults that span, supported by iron beams, from south to north across the plan (Figure 8.3).

The floor is made of clay tiles with inserts of brick. The church is entered at the southwest corner through a door that is protected by a tall, freestanding brick screen. The plan consists of a nave with a single "aisle"

Figure 8.3 Sigurd Lewerentz, St. Mark's Church, Björkhagen, Stockholm. Interior.

to the north. Here are the baptismal font, carved from a single block of North Swedish limestone, and the organ, housed in a timber case designed by Lewerentz.

Against obvious expectation, Lewerentz compounds the dark materiality of the interior by casting only minimal quantities of daylight upon it. The body of the nave is lit through only five openings, all in the south wall. The largest are a pair of windows, separated by a narrow brick pier, that rise from floor level to illuminate the sanctuary. To the west of these are two square openings with high sills and high in the wall close to the entrance is a window that is shaded by the curved brick screen at the entrance. The "aisle" is lit by two slots of light that are formed by the positions where the north wall breaks in echelon, casting oblique east light at "grazing" incidence along the rich texture of the brickwork.

The outcome of this configuration of volume, material and light is a complex field of luminance that highlights specific points in the space, the altar table, the pulpit, font and organ within the low ambience of the whole. Lewerentz's use of frameless sheets of glass, attached to the inner face of the brickwork of the south wall, emphasises the contrast of brightness between outside and in. On entering the church the darkness is first tempered by the presence of the arrays of polished copper and brass light fittings that float in the space. These provide artificial illumination to supplement that from the windows, but their specular surfaces also pick up and inter-reflect light from the windows, adding an additional element to the visual field. Small candelabra and carefully sited spotlights bring yet more diversity of light, as do the glittering gilt and silver of the cross and candle-sticks that adorn the altar.

In the north aisle artificial light is used to illuminate both the baptistry and the organ. A glittering brass chandelier hangs above the font, drawing the eye to this significant locus of the church and symbolising the importance of the act of baptism. The timber organ case, in effect a mini-ature building-within-a-building, receives general light from the high, east-facing window, but is itself a source of light, when the bright light from its music stand floods out into the body of the building.

A particular property of visual adaptation is that the general level of light to which we are adapted determines how and what we see. This means that, at low levels of ambient light, specific objects appear brighter than they would in brighter surroundings. It is precisely this effect that Lewerentz exploits at St. Mark's. In the general darkness of the interior the intense patches of natural and artificial light stand out in ever shifting rela-tionships. The space is not merely lit, but becomes a composition of lights. The other significant property of adaptation is the process by which we gradually become adapted to low levels of light. As Gregory explains, so-called "dark" adaptation takes place in the first few minutes of dark-ness, but rod and cone cells adapt at different rates. The cones complete their adaptation in about seven minutes, but rod cells continue to adjust for over an hour. This means that the worshipper at St. Mark's will gradually perceive more detail of the space as the church service progresses, a powerful metaphor of revelation.

The undisguised raw brick construction of the church led Lewerentz to invent a method for the suspension of the light fittings in which cables span across the space to carry the individual lamps and their electricity supply. Conduits, cables and switches are fixed to the brickwork at just the places they are needed, not as clumsy afterthought, but as essential expression of the materiality of the building. But, whilst these services are so explicitly displayed, Lewerentz was discreet about heating and ventilation systems. The cavities of the thick enclosing walls of the church accommodate a forced air system that discharges warmth through clusters of open perpends in the brickwork, particularly in the south wall of the nave. These small voids also perform an acoustic function by acting as absorbent cavities to control the reverberation of the otherwise highly reflectant materials. The undulating rhythm of the brick vault will also help the acoustics by diffusing reflected sound.

Lewerentz's sensitivity to the nature of light is demonstrated at St. Mark's by the manner in which he fused natural and artificial sources into a rich and original unity. His understanding of the mechanism of visual adaptation, almost certainly intuitive rather than scientific, allows a particular kind of vision, both literal and metaphorical, in the darkness of the church that reflects and sustains the mystery of religious ritual.

St. Peter's Church, Klippan

St. Peter's at Klippan near Helsingborg followed soon after St. Mark's. It, self-evidently, continues the themes first explored at Björkhagen, and most critical opinion regards it as an intensification of these. The plan is, again, extremely simple, but more condensed, with the rectangular block of the church held within the angle of an L-shaped building that houses the parish functions. The church is a precise square in plan, with two small wings projecting from its north side. Unlike the aisled plan of St. Mark's, the church is a single volume. A brick vaulted roof, similar to that at Björkhagen, is now oriented east to west and the span is reduced by a secondary steel structure – although that is hardly an appropriate term. Pairs of steel sections make up a T-shaped frame that stands at the centre of the space. This supports two transverse beams that carry the vaults. The whole enclosure is made of dark Helsingborg brick, now also used for the floor, which slopes gently from west to east, and is penetrated by even smaller openings than at St. Mark's.

The square windows, two in the west façade and two facing south, are, in detail, an inversion of those at St. Mark's. Here the glass is fixed to the external face of the wall with a simple detail of steel plates and mastic pointing. This shift brings the deep brick reveals within the interior and conveys a more acute sense of the enclosing presence of the walls. But, unlike St. Mark's, windows are not the only source of natural light to be found in the church. In another tectonic invention, perhaps even more remarkable than the window detail, Lewerentz constructs a series of roof-light shafts that rise above and illuminate key places. Two of these are in the church; elsewhere they bring light to the wedding chapel and waiting room and to the sacristy. Apparently similar at first sight, they are subtly differentiated in their orientation and materiality.

The relation of windows and rooflights to the volume and material means that the interior of the church is even darker than St. Mark's. The sequence of entry leads from the north, between the projecting wings of the wedding chapel and the belfry. The actual entrance lies through the dimly lit wedding chapel and here begins the process of adaptation that prepares the eye for the broad vista of the church, first seen from the southwest corner. This threshold is illuminated by the paired, square windows in the west wall immediately to the right. These cast light onto the remarkable baptismal font, just to the left, with its giant, white, reflectant seashell suspended, on a black metal frame, over a fissure in the brick floor. Looking beyond the font the visual field is complex. Here walls, roof and floor are all of Helsingborg brick, against which pools of daylight sparkle from the south-facing windows and are projected down from the rooflights (Figure 8.4).

Mapped over this tectonic background, the artificial lighting uses the same polished copper and brass lamps as St. Mark's, emitting and reflecting multiple light sources as before. At key points on the east and

Figure 8.4 Sigurd
Lewerentz, St. Peter's
Church, Klippan. Interior.

north walls, pinpoints of naked light from candelabras flicker against the brickwork. The pair of rooflights above the church are glazed horizontally. They are oriented north–south and capture the brightest light of the high sky, projecting it precisely onto the priest's path from the sacristy to the sanctuary signalling his entrance and the beginning of the service. The orientation means that direct sunlight enters the space at midday. Ahlin suggests that the image of slanting shafts of light from above, which he calls "bundles of light," was drawn by Lewerentz from the old brick factory at Helsingborg; the source of the bricks for both St. Mark's and St. Peter's.[13]

Light and lighting, darkness and shadow were essential elements of Lewerentz's work from the earliest projects. At St. Peter's these are brought together in a complex composition that is progressively revealed as the eye adapts and as the light itself changes with the passage of time. In contrast to the brightness that prevails in so much architecture of modern times, the literal and metaphorical darkness of this building uncannily makes the nature of light more apparent.

Lewerentz uses a system similar to that adopted at St. Mark's to support and supply the hanging lamps. But here the supporting cables pass from east to west over the intervening steel structure. The space between the sloping brick floor and the brick vaulted roof is thereby vertically layered, primarily by the steel structure, then by the web of dark wires and cables and the polished metal lamps that they support.

The heating system is also a development of that used at St. Mark's. A basement plant room beneath the parish block houses the boiler and air-handling units from which air is ducted to the church. Air is supplied through ducts to open joints within the brickwork of the walls. Return air is collected high on the north wall. The warmth to be encountered within the building is visually anticipated by the mass of the brick flue, capped by three large chimney pots that act as a pivotal element in the composition of the space between the church and the parish building.

Acoustics are the final element of the environmental ensemble at St. Peter's. As we found at St. Mark's the darkness of the church is complemented by a calm acoustic that, at first sound, seems at odds with the apparent hardness of the brickwork and the volume of the space. This calm is the result of the acoustic absorption of the cavities in the brickwork and of the diffusing effect of the roof vault. Most important of all is the intermittent sound of baptismal water as it drips from the shell font into the fissure in the brick floor. Quietly insistent, this sound reinforces the inward-ness of the church and clinches its separation from the everyday world without.

These churches, with their knowing tectonics and sophisticated services installations, strive to reach a profound representation of the human environment. Most particularly they teach us that this environment is a complex overlapping of sensations, of natural elements, particularly of light, and mechanical provisions, of all-enveloping warmth in the cold of the northern winter, and of subtle acoustics. Quantitatively and qualitatively these sensations change almost imperceptibly as time passes and as our bodies and minds assimilate and respond to the rich ambience of the spaces that Lewerentz created.

Churches of Zumthor and Siza

In the churches of St. Mark at Björkhagen and St. Peter at Klippan, Lewerentz realised a remarkable synthesis of the elements of architecture: *form, material* and *environment*, that was without precedent and, even though these buildings are widely admired, without direct progeny. They stand as the products of an original imagination, enigmatic and provoca-tive. But, most significantly, they declare the potentiality of the methods of modern architecture to give expression to the sacred. The two more recent churches by, respectively, Peter Zumthor, at the Chapel of St. Benedict at Sumvigt in Switzerland (1985–1988) and Alvaro Siza, in his design of the church of Santa Maria at Marco de Canavezes in Portugal (1990–1996), are further instances of this.

St. Benedict's Chapel, Sumvigt, Switzerland, 1985–1988

At the heart of Peter Zumthor's work lies a deep engagement with architecture's capability to engage the senses and emotions.

> [T]he quest for the new object that I shall design and build consists largely of reflection upon the way we really experience the many

places of our so different dwellings throughout the world – in a forest, on a bridge, on a town square, in a house, in a room, in my room, in your room, in summer in the morning, at twilight, in the rain. I hear the sounds of cars moving outside, the voices of the birds, and the steps of the passers-by. I see the rusty metal of the door, the blue of the hills in the background, the shimmer of the air over the asphalt. I feel the warmth reflected by the wall behind me. The curtains in the slender window recesses move gently in the breeze, and the air smells damp from yesterday's rain, preserved by the soil in the plant troughs.

(Peter Zumthor)[14]

With these perceptions in mind the tiny chapel of St. Benedict at Sumvigt offers a wonderful opportunity to explore the environmental imagination. Its very situation high on the northern slopes of the upper Rhine Valley is the starting point. Its scale and materiality take cues from the scattered houses of this mountain community, but its form and detail declare its special purpose. It seems to be inevitably of the place.

Every new work of architecture intervenes in a specific historical situation. It is essential to the quality of the intervention that the new building should embrace qualities which can enter into a meaningful dialogue with the existing situation. For if the intervention is to find its place, it must make us see what already exists in a new light. We throw a stone into the water. Sand swirls up and settles again. The stir was necessary. The stone has found its place. But the pond is no longer the same.[15]

On first encounter the building might be considered to be more about the tectonic than the environmental, but Zumthor has insisted that such conventional categories and distinctions have little relevance for him.

The sense that I try to instil into materials is beyond all rules of composition, and their tangibility, smell and acoustic qualities are merely elements of the language that we are compelled to use. Sense emerges when I succeed in bringing out the specific meanings of certain materials in my buildings, meanings which can only be perceived in just this way in this one building.[16]

The first impression within the chapel is of simplicity. This is conveyed by the combination of clear form, repetitive and comprehensible structure illuminated by the continuous band of the clerestory window (Figure 8.5).

The conventional east–west orientation of the Christian church renders this configuration asymmetrical through the play of light, more intense from the south than the north. Although the interior is brightly lit and all is literally visible, the richness and complexity of the building progressively comes "into view" in a way that is analogous to the process of visual adaptation.

The dully silvered finish of the plywood lining of the walls attracts the light from the clerestory and, by inter-reflection, animates the space. A specific artefact of this process is the intensification of brightness that

Figure 8.5 Peter Zumthor, St. Benedict's Chapel, Sumvigt. Interior.

occurs by inter-reflection between the natural timber of the structural columns and the silver wall. Another detail that comes to the attention is the fine tapering of the mullions of the clerestory. This, in the manner of the moulded glazing bars of old window frames, softens the contrast between mullion and sky and thereby avoids glare. As the sun follows its daily course across the sky the space is animated by the ever-changing play of sunlight on the interior. Once again the simplicity of the continuous clerestory, in its relationship to the non-orthogonal plan, produces effects of surprising richness.

Zumthor has proposed that

> In architecture, there are two basic possibilities of spatial compo-
> sition: the closed architectural body which isolates space within
> itself, and the open body which embraces an area of space that
> is connected with the endless continuum . . . Buildings which
> have a strong impact always convey an intense feeling of their
> spatial quality. They embrace the mysterious void which we call
> space and make it vibrate.[17]

St. Benedict belongs to the first of these categories. There is an intense focus upon the "space within itself" that is the almost inevitable effect of the absence of windows at eye level. This effect is rendered particularly powerful by the minute scale of the chapel. But there is simultaneously a sense of the vast scale of the terrain, the "endless continuum," outside, promoted by the view of the sky seen through the clerestory.

Acoustically the building has the particular qualities that follow from the use of light timber construction. It is responsive to human presence, sounding every footfall, and there is an audible awareness of the almost imperceptible movement of the structure with gentle creaks. One is aware of bird song and the sound of the wind in the trees of the mountainside. All of which is expressive of the nature of the building itself and of its relationship with its site.

High in the mountains the building is exposed to extreme winter weather. The response to this couldn't be simpler. The timber construction permits a high standard of thermal insulation to be achieved within the structure and simple electrical heating elements concealed beneath the pews produce warmth exactly where it is needed. The only other "service" is the array of electric lamps that hang from the roof, following the plan form, to provide nighttime illumination. In their simplicity these are absolutely in accord with the whole conception of the building.

Santa Maria, Marco de Canavezes, Portugal, 1990–1996

> I wanted to make a church that felt like a church and not a
> building with a cross in it. I wasn't interested in this primitive

notion of how a symbol could determine the character of a building. So I tried to achieve something I would call the character of the church . . .

(Alvaro Siza)[18]

The church is located just outside the centre of the town of Marco de Canavezes, in an area of undistinguished modern development. It sits on a podium above a dual carriageway road. This elevation above its surroundings enables it to establish a strong presence in this nondescript context. The podium is constructed from the local granite. The white, orthogonal volume of the church rises from the podium and a granite dado of variable height negotiates the transition of material. The church is a simple rectangle. To the west two projections either side of the grand entrance contain, to the north, the baptistry and the belfry to the south. Within this simple formal scheme spatial complexity is achieved through manipulation of the details of plan and section.

The essence of this is the asymmetry between the north and south faces. To the south the tall vertical wall is penetrated only by a low, horizontal window.

The north wall is a leaning, convex surface that terminates in three large clerestory windows. At the east it intersects the convex form of the nave at the point where it hovers above the side chapel. The interior is predominantly white with all the walls and the ceiling finished with matte paint (Figure 8.6). The west, south and east walls have a dado of white tiles, whereas the north wall is painted over its full height. The main area of the floor and that of the sanctuary are of wide boarded hardwood, whereas the rear of the nave, the baptistry and the bell tower are floored with white marble. The principal light of the space is from the north-facing clerestory. The lower, horizontal window in the south wall illuminates the territory close to it, and will admit direct sunlight at certain times of day. The north wall itself is primarily illuminated by light reflected back from the clear, blank surface of the south wall.

Looking west, the tile-lined baptistry and, to a lesser extent, the belfry are the most brightly illuminated parts of the entire composition. The overall effect is calm, but full of subtle differentiations as the forms and materials respond to the predominantly shadowless illuminance. This also contributes to a sense of thermal comfort that is the outcome of the volume, its heavyweight construction and the absence of direct sunlight. On the hot summer's day it was very cool at midday. A further, subtle contribution to the sense of calm and cool, reminiscent of Lewerentz at Klippan, is provided by the sound of running water that is instantly apparent on entering the building. This comes from the baptistry, where water constantly issues from the font into the recessed stone basin in which it sits. The acoustic effect of this is accentuated by the materiality and sheer height of the baptistry tower.

Figure 8.6 Alvaro Siza, Santa Maria, Marco de Canavezes. Interior.

This is a building that gives precedence to "atmosphere" over the tectonic. The materiality of the building is primarily of surface – white-painted plaster, white tile, white marble, timber, granite – than it is of expressed construction. In a Kahnian sense, light is perceived as a building material. The subtle differentiations of brightness that play within the volume of the church invest it with a rich calm (if that is not a contradiction in terms) – serenity. The three physical parameters of environment, heat, light and sound all have a role in establishing the atmosphere – low, calm light, cool, still air and the quietly audible sound of running water.

All of this works to dissociate the interior from the untidy bustle of the town outside. This effect is reinforced by the visual control that is achieved by the horizontal window in the south façade. This directs the attention towards the hilltops of the distant horizon rather than to the road and nondescript modern urban buildings in the foreground.

In the European tradition, the alternative schools of Gothic and classic dominate Christian architecture. Zumthor's and Siza's buildings might be said to represent a contemporary continuation of that distinction. The Chapel of St. Benedict is a refined essay in expressive construction and at Santa Maria materiality is concealed behind an almost uniform, applied surface of "white" – inside and out. Each position has a fundamental and profound effect on the respective building's environmental qualities.

St. Benedict manifestly offers shelter from the elements of the often-harsh climate in which it is set. Its copper roof and sheathing of shingles speak of enclosure and protection. Inside, Zumthor arranges structure, material and light so that their interaction creates a rich and complex setting for worship. The apparatus of modern environmental systems is reduced to the expression of the simplest of light fittings and the heating system is simply and artfully concealed from view.

The cubic, white volume of Siza's Santa Maria, with its roof concealed by a parapet, is less evidently concerned with the question of shelter. The uniform whiteness of the interior allows subtle gradations of light to be visible and serves to invest the space with an appropriate calm. Once again the devices of service systems are absent from view to sustain the purity of the white space.

What is common to these buildings is environmental *imagination*. This is the ability to envision the outcome of the conjunction of form and material, set within the physical facts of the climate and locale, in ways that inform and enhance the purpose and meaning of a building.

Airs, Waters, Places[19] – Peter Zumthor, Therme Vals: Body and Environment

In Victor Olgyay's, *Design with Climate*[20] there is one of the most wonderful images of modern architectural science. A man (looking, from the rear, uncannily like Ronald Reagan) stands exposed to all of the conceivable means by which he might exchange heat with his surroundings (Figure 8.7).

His body produces heat by the metabolic processes of human physiology (1a–d). He absorbs radiation from the sun, from glowing radiators and from non-glowing objects and surfaces (2a–c). Heat is conducted to his body from the surrounding air, if warm, and by contact with surfaces (3a–b). He is affected by the condensation of atmospheric moisture (4). The man then may lose heat by radiation to the sky – if it is cool – and to

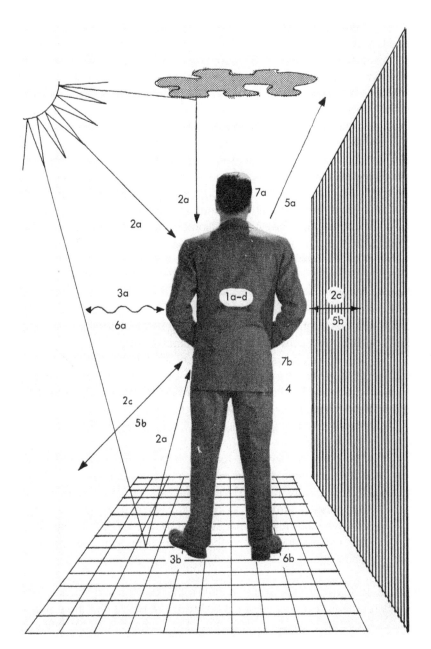

Figure 8.7 Heat exchange between man and surroundings. After Victor Olgyay, *Design with Climate*.

cold surfaces in his surroundings (5a–b). Heat may be conducted away to surrounding cool air and to any cool surfaces with which he is in contact (6a–b). Finally, heat may be lost by evaporation through the respiratory tract or from the skin (7a–b). In the image all of these processes are depicted operating simultaneously. Thankfully this circumstance would be inconceivable in any practical environment except, perhaps and unspeakably, a torture chamber.

I use this image, and its accompanying commentary, to demonstrate that thermal environments, even those that we define as "comfortable" are complex in combining in some measure processes of conduction, convection, evaporation and radiation. When we extend our terms of reference to include the luminous and acoustic aspects of building environments, in all of their potential diversity, we can see that, even as we go about our most ordinary daily business, we are immersed in an elaborate environmental cocktail. In order to meet conventional codified notions of "comfort" we find that practical environments tend towards the middle ranges, neither too warm or too cool, too bright or too dull, too loud or too quiet. On the other hand, the most memorable and remarkable architectural environments often break through the bounds of convention. They discover combinations of the environmental elements that, by some particular emphasis or relationship, enrich the experience of inhabitation.

Therme Vals

This building by Peter Zumthor, completed in 1996, embraces all of the elements of heat, light and sound, in distributions and combinations that extend beyond most conventional circumstances.

Above all this is an environment of the senses. Unlike the conventions represented by Victor Olgyay's besuited man, where contact with the environment is limited and decorous, at Vals we enter an altogether different and infinitely more sensuous condition.

In describing its essential nature Zumthor has written:

> The building takes the form of a large, grass-covered stone object set deep into the mountain and dovetailed into its flank. It is a solitary building, which resists formal integration with the existing structure in order to evoke more clearly – and achieve more fully – what seemed to us a more important role: the establishing of a special relationship with the mountain landscape . . .[21]

It is this acute awareness of the mountain that defines and characterises the environmental experience of the building. Examination of the plan and the cross-sections immediately conveys an understanding of the intimacy and richness of the relationship of the construction to the topography and geology of the valley (Figure 8.8). From the plan we may note how the density of the building, its "geological" formation, changes as it unfolds from its innermost depths, by the central pool with its surrounding monoliths of the load-bearing "stones," and towards the alternation of solid and void that constitutes the eastern edge. We may also notice how the topography of the building opens towards the south, terminating in the open-air pool and its terraces. The cross-section reveals the "cut-and-fill" of the

Figure 8.8 Peter Zumthor, Therme Vals. Exterior.

building into the mountain slope. These properties directly influence the environmental experience of the building. As a bather you arrive at the northwestern corner of the plan deep in the mountain and progress through ever-changing environments towards the more open eastern and southern extremities. You first pass through exquisite, boudoir-like, polished hardwood, changing rooms and emerge on a gallery above a stepped ramp that leads towards the internal pool. On the downward journey you have, first, a view across the central pool then, at the foot of the ramp, as you turn eastwards, you are presented with a framed view of the opposite hillside. From here you explore at will the variety of sensory experiences that the building offers.[22]

Experience and Environment

The primary experience of the spa is of bodily immersion in the waters, with their diverse temperatures, but over and above this the building offers a complex synthesis of sensory stimuli. The bather encounters, in endless combinations, the atmosphere, or *aura* as Friedrich Achleitner has insisted,[23] its temperature, humidity, luminosity, scent and the sound that it carries.

Visually the environment ranges from the wholly artificial light of the spaces within the "stones"; the glittering dark of the interior zones, with orchestrated brightnesses of the narrow shafts of light that enter with surprising brightness between the roof slabs and small hanging lamps, whose light is more absorbed than reflected by the striated surfaces of gneiss stone; the iridescent blue of the array of rooflights above the central pool, the bright daylight of the eastern gallery and, finally, the full brightness of the sky bounded by the horizons of the mountains rising above the outdoor pool.

The water that flows from the mountain spring at Vals is at a temperature of 30°C and it is water and its temperature that define the thermal environment of the building. Within the spa the central pool is at 32°C. The outdoor pool is at 36°C, to compensate for exposure to the ambient temperature. Elsewhere, as voids within the stone masses are discovered, the bather may choose from a variety of water temperatures, contrasting the heat of the fire bath at 42°C with the 14°C chill of the nearby cold bath. The "flower" bath is at a gentle 30°C and offers olfactory pleasure as gentle scents rise from the jasmine petals that float on the surface of the water. The successive chambers of the Turkish bath offer a yet more extreme thermal experience that is emphasised further by the dim illumination, the swirling, steam-filled, heavily scented atmosphere.

Throughout the spa the acoustic is as diverse and as subtle as the waters. In a conversation with Steven Spier, Zumthor has stated that "I believe that buildings should sound the way they look."[24] This seemingly simple statement conceals a subtle and original grasp of environmental relationships. In purely physical terms the acoustic of a space is the product of its geometry and its materiality. It is, in the reductive terms of building science, a function of volume and sound absorption.[25] The four principal materials that define the internal spaces of the spa, stone, concrete, glass and water, all offer very little absorption of sound.

Nonetheless Zumthor's manipulation of orthogonal geometry establishes many different relationships of volume and material that influence the perceived sounds of the building. By some almost magical process the building seems to quieten its users. Conversation is hushed as bathers move about the building and discover and respond to its acoustics, so that the impression is quite unlike the reverberant clatter of a conventional indoor swimming pool. To rephrase Zumthor's statement about the sound of buildings, the look of the building tells you what sound to make, or indeed in some instances to be silent. Perhaps the most striking instance is set deeply into the northwest corner, below the point of entrance, where an almost concealed passage leads to the "spring grotto." In this tall volume of rough-faced stone a unique acoustic effect is created. The echoing reverberation of the space somehow provokes bathers to hum or sing as they sense the acoustic. In the outdoor pool, where the acoustic is that of the valley rather than of the building, the force of three water jets creates a continuous percussive rhythm as it strikes either the surface of the pool or the backs of bathers. All of these effects are the product of "natural" acoustics; the bather "plays" building almost as a musician plays an instrument. In one unique place, the "sounding stone," you encounter the "artificial" sounds of a recorded composition specifically written for the building by the composer/percussionist Fritz Hauser. Here you are the receiver rather than the producer of sound, passive rather than active.

Zumthor often speaks of his interest in music:

> [T]he slow movements of the Mozart piano concertos, John Coltrane's ballads, or the sound of the human voice in certain songs all move me.
>
> The human ability to invent melodies, harmonies and rhythms amazes me.
>
> But the world of sound also embraces the opposite of melody, harmony and rhythm. There is disharmony and broken rhythm, fragments and clusters of sound, and there is also the purely functional sound we call noise. Contemporary music works with these elements.[26]

He has also noted the relevance for architectural thought of John Cage's compositional process where

> . . . he is not a composer who hears music in his mind and then attempts to write it down. He has another way of operating. He works out concepts and structures and then has them performed to find out how they sound.[27]

The analogy of architecture and music can be dangerously misleading, often banal, but it seems to be useful in attempting to describe something of the sonic essence of Therme Vals. In his conversation with Steven Spier,[28] Zumthor tacitly acknowledged that his approach to the acoustics of the building – and to its lighting – rejected conventional normative prescriptions in favour of reference to "a personal body of experience." In the light of this it is possible to represent the *sound* of Vals by analogy with Zumthor's characterisation above of the nature of contemporary music and his reference to Cage's compositional method. The acoustic experience is of "fragments and clusters of sound" as speech, footfall and the sound of rippling water are inter-reflected about and within the complex spatial organisation of the building.

By its nature architecture, particularly architecture as concrete as that at Vals, can only partially subscribe to the indeterminacy of Cage's music. Working with the facts of material and volume the acoustic outcome is inevitable, could be a matter of calculation, and when constructed be almost certainly beyond modification. It is here that Zumthor's resort to "a personal body of experience" comes into play. As he told Spier:

> I have to get into all the possible qualities which could be brought, which arise within me, out of my memory, experiences, fantasies and images, to generate this building. And I do this without any programmatic ideas in my head. . . . The way I have been brought up helps me to start really independently from rules, books, and things, so that I can try to be true to what I feel.[29]

Just as Cage's music is free, but not arbitrary, Zumthor's architectural method is supported by the security of memory and experience, one might call this "informed intuition," in the search for the solution. The unique and complex sounds of Therme Vals are not the result of calculation and analysis. How could such aims have been codified? But they are the result of a process in which elements of memory and experience are brought to bear on the qualities of the evolving design.

Day to Night

The transition from day to night brings a further dimension to the environmental experience at Vals. This is most immediately transmitted through the medium of light, as the dynamic directional flow of daylight from the east and south is replaced by more static artificial sources within. That is, of course, the condition of all modern buildings, but there are few where the transformation is quite as remarkable as here.

Simple metal-shaded pendant lamps supply the background lighting of the circulation and poolside areas, but the primary illumination comes from the pools themselves. Recessed lamps, set below the water line, cast arcs of light through the water and these, in turn, softly illuminate the surrounding stone walls. In the central pool this light just reaches the concrete surface of the ceiling, but here the 16 blue glass rooflights are themselves bright, lit by external spotlights. In the outdoor pool the glowing surface of the water itself becomes one of the most magical luminaries. Its light is carried upwards by the steam rising from the water creating, particularly in winter, a territory of light and warmth beneath the dark and cold of the sky. This magical quality evoked by light and water is reinforced by the absence of sound. In the evenings the water jets in the outdoor pool and Fritz Hauser's recorded music in the sounding stone are silenced and the whole building achieves a heightened calm. Bathers swim slowly and speak in hushed tones.

As I have tried to show in describing Therme Vals, the building offers a rich, complex and entirely sensual experience. One's body and spirit are immersed in a combination of atmospheres, hot and cold, light and dark, quiet and reverberant, perfumed, tactile – airs, waters, places. These are sustained by technical expertise of the highest order, but this is, in terms of Zumthor's "talk" metaphor, silent.

Conclusion

Of necessity modern buildings use the tools of quantified norms and mathematical calculation to deliver practical environments for diverse functions. This is a matter of *technics* and a vital part of practice. But over and beyond this lies the potential to fashion environments that reach beyond the practical and enter the realm of *poetics*. This is achieved in each of the buildings that I have described and is the outcome of a process of conception and realisation that is quite different from but just as disciplined as that of engineering practices. It is my hope that the studies presented in this essay capture and communicate something of the essence of this process.

The research method adopts and adapts approaches from both architectural science and the humanities and might be productively followed in future studies. My selection of architects and their works has, inevitably as in all field-based research, been constrained by time and material resources. But in my mind there are numerous other subjects that, in my view, would repay investigation by these methods. For example, studies of Schinkel and Semper would add much to the account of events in the nineteenth century and architects such as Wright, Loos, Terragni, Barragan, Utzon and Scharoun are conspicuously absent from my studies of the twentieth century. Many architects currently practising seem to me to be adding further richness to environmental experience in architecture. Amongst these, Ando, Botta and Murcutt come immediately to mind. A further fruitful line might be to explore works that share a common geographical setting. In particular, I am aware that my studies so far have paid relatively little attention to the special environmental qualities of British architecture. I have an intuition that a common thread may run through British work from the beginning of the nineteenth century to the present day, as architects have sought to express the nature and meaning of the particular environmental conditions of the British Isles.[30] There is much to do and, perhaps, this short essay will stimulate others to bring other insights to bear on the poetics of environment.

Notes

1. Dean Hawkes, *The Environmental Tradition: studies in the architecture of environment*, Spon Press, London, 1996.

2. The author's completed research project was supported by an Emeritus Research Fellowship awarded by the Leverhulme Trust, 2002–2003. A full account of the project is published in Dean Hawkes, *The Environmental Imagination: Essays on the Poetics of the Architectural Environment*, Spon Press, London, 2006.

3. Steen Eiler Rasmussen's *Experiencing Architecture*, MIT Press, Cambridge, MA, 1958, is, perhaps, the earliest modern research to address the question of environmental *experience*. Juhani Pallasmaa's *The Eyes of the Skin: Architecture and the Senses*, Academy Editions, London, 1996, eloquently explores the complexity of our sensory mechanisms, as they encounter the conditions of architecture.

4. Francesco Dal Co, "The Architecture of Carlo Scarpa," in Francesco Dal Co and Giuseppe Mazzariol, *Carl Scarpa: The Complete Works*, Milan/London, Electa/the Architectural Press, 1986.

5. Carlo Scarpa, I wished I could frame the blue of the sky, from a recording of a lecture given on 13 January 1976, in *Rassengna, Carlo Scarpa, Frammenti, 1926/78*, 7 June 1981.

6. Sergio Los, "Carlo Scarpa – Architect and Poet," in *ptah: architecture design art*, 2001, 2, pp. 59–77.

7. Sergio Los, ibid.

8. Sergio Los, ibid.

9. Carlo Scarpa, op. cit.

10. See Richard Gregory, *Eye and Brain: the psychology of seeing*, Weidenfeld and Nicolson, London, 1966, for a clear exposition of the physiological and psychological bases of adaptation.

11. Juhani Pallasmaa, op. cit.

12. For comprehensive descriptions of Lewerentz's works see Janne Ahlin, *Sigurd Lewerentz: architect*, Byggförlaget, Stockholm, 1985, English edition, Byggförlaget/MIT Press, Stockholm and Cambridge, MA 1987 and Nicola Flora, Paolo Giardiello and Gennaro Postiglione (eds.), *Sigurd Lewerentz; 1885 1975*, Electa, Milano, 2001.

13. Janne Ahlin, op. cit.

14. Peter Zumthor, *Peter Zumthor Works: Buildings and Projects, 1979–1997*, Lars Müller Publishers, Baden, Switzerland, 1998.

15. Peter Zumthor, "A Way of Looking at Things," in *Thinking Architecture*, Birkhäuser, Basel, Boston, Berlin, 1999.

16. Peter Zumthor, ibid.

17. Peter Zumthor, ibid.

18. Alvaro Siza, interview with Yoshio Futagawa in *Alvaro Siza, GA Document Extra*, no. 11, 1998.

19. *Airs, Waters, Places* is the title of the collection of medical writings known as the *Hippocratic Corpus*. These were written between 430 and 330 BC by Hippocrates and other authors. *Airs, Waters, Places* is "an essay on the influence of climate, water supply and situation on health." In modern terminology it is about what is now referred to as "environmental health" and bears much likeness to Vitruvius' discussion of the siting of a city, *Ten Books on Architecture*, Book 1, Chapter 4, M.H. Morgan (trans.), Dover, New York, 1960. The title has been appropriated for the present essay because it captures the essence of Peter Zumthor's building, both in its physical nature as a rich complex of environmental experiences and in its Hippocratic association with human well-being. The substance of the Hippocratic Corpus is comprehensively presented and discussed in G.E.R. Lloyd (ed.), *Hippocratic Writings*, Penguin Classics, Penguin Books, London, 1978.

20. Victor Olgyay, *Design with Climate: Bioclimatic Approach to Architectural Regionalism*, Princeton University Press, Princeton, NJ, 1963.

21. Peter Zumthor, *Peter Zumthor Works: Buildings and Projects 1979–1997*, op. cit.
22. It is a principle of Therme Vals that the bather should be free to decide how to use the spa, in what sequence and for how long to enjoy the various pools and terraces.
23. Friedrich Achleitner, "Questioning the Modern Movement," in *Architecture and urbanism: Peter Zumthor*, Extra Edition, February, a+u Publishing, Tokyo, 1998. Achleitner writes: "Even if Peter Zumthor doesn't like it, the word *aura* is useful here. *Atmosphere* is not enough."
24. Steven Spier, "Place, authorship and the concrete: three conversations with Peter Zumthor," in *arq (architectural research quarterly)*, vol. 5, no. 1, 2001, pp. 15–36.
25. This relationship was first objectively described by W.C. Sabine, the founder of modern architectural acoustics, in his essay, "Reverberation," in *The American Architect*, 1900, reprinted in F. Hunt (ed.), *Collected Papers on Acoustics*, Dover, New York, 1964.
26. Peter Zumthor, "A way of looking at things," in *Thinking Architecture*, op. cit.
27. Peter Zumthor, 'The hard core of beauty', in *Thinking Architecture,* op cit.
28. Steven Spier, op. cit.
29. Ibid.
30. This project has been addressed to some degree in my recent book, *Architecture and Climate: an environmental history of British Architecture 1600–2000*, Routledge, London, 2012.

Chapter 9

Why does the Environment Matter?

Derek Clements-Croome

Introduction

For an organisation to be successful and to meet its necessary business targets, the performance expressed by the productivity of its employees is of vital importance. In many occupations people work closely with computers within an organisation, which is usually housed in a building. Today, technology allows people to work easily while they are travelling, or at home, and this goes some way to improving productivity. There are still, however, many people who have a regular workplace, which demarcates the volume of space for private work but is linked to other workplaces as well as to social and public spaces. People produce less when they are tired; have personal worries; suffer stress from dissatisfaction with the job or the organisation. The physical environment can enhance one's work, but an unsatisfactory environment can hinder work output.

Concentration of the mind is vital for good work performance. Absolute alertness and attention are essential if one is to concentrate. There is some personal discipline involved in attaining and maintaining concentration, but again the environment can be conducive to this by affecting one's mood or frame of mind; however, it can also be distracting and can contribute to a loss of concentration.

A number of personal factors that depend on the physical and mental health of an individual, and a number of external factors that depend on the environment and work-related systems, influence the level of productivity. Experimental work on comfort often looks at the responses of a group as a whole and this tends to mask the individuals' need for sympathetic surroundings to work and live in. People also need to have a fair degree of personal control of various factors in their environment. They react to the environment as a whole, not as discrete parts, unless a particular aspect is taxing the sensory system.

Productivity can be measured in absolute or direct terms by measuring the speed of working and the accuracy of outputs by designing very controlled experiments with well-focused tests. Comparative or indirect measures use scale and questionnaires to assess the individual opinions of people concerning their work and environment. Combined measures can also be employed, using for example some physiological measure such as brain rhythms to see whether variations in the patterns of the brain responses correlate with the responses assessed by questionnaires.

Environment, Mind and Brain

From the moment of conception, a life is generated that physically develops in the surroundings of the womb. The life of the mother – what she eats, how she lives, her environment – all leave their imprint on the developing embryo. For example, smoking and alcohol should be avoided. Some people believe that oxygen purging helps to develop the brain; others recommend that birth should take place in hydro pools. These things indicate the importance of the environment, which is evident at the earliest stages of life. A continual life process of environmental learning has begun. At birth the connections linking brain cells are few, but rapidly these grow and multiply as experiences of environment are felt each moment in the conscious state (Greenfield 2000). During sleep experiences such as dreams still occur, but there is a threshold shift to limit the impact of environmental stimuli. Our genetic imprint means we respond individually to environmental stimuli, but in the case of extreme conditions human reactions tend to converge. Our genetic imprint in any generation is a result of a gradual evolution, which itself has stemmed from historic genetic roots, which in turn have been influenced by the environment. The environment plays an important part in this evolutionary process.

The brain has several systems within it that influence the human reaction to the environment. The *limbic system* interacts with the cerebral cortex to pick up perceptions and memories and uses the facilities of the cortex for analysing such data. The *hippocampus* plays a part in evaluating the incoming stimuli in terms of one's past experiences. The *amygdala* functions to intensify an emotional response whenever the incoming stimuli do not fit the expected patterns. Another part of the limbic system is the *septal region*, which plays a part in toning down our emotional reactions.

A vital part of the brain is the *reticular activating system*, which decides from one moment to the next what incoming sensory information, if any, should enter our consciousness. It acts as a gatekeeper to our consciousness and in so doing determines the world that we perceive.

A sensation occurs when an organ or sense is stimulated, whilst the perception depends upon the sensation, but also includes a conception

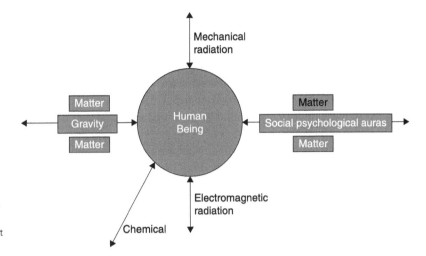

Figure 9.1 Components of self and its relationships with the body, mind, spirit and soul.

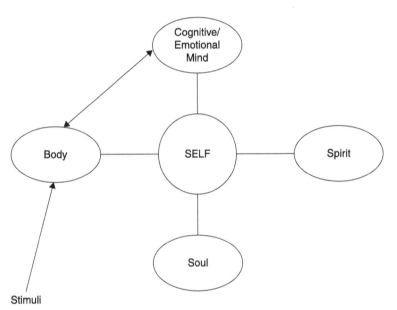

Figure 9.2 Relationships between the human being, the environment and matter.

of the object perceived. Put in another way, sensation is a process of detecting a stimulus or some aspect of it in the environment, whilst perception is the process by which we interpret the information gathered and processed by the senses.

What is happening to let us experience events? Environmental stimuli arise from our sensory system interacting with the world around us such as other living things, our occupation at any moment and the solid and fluid world around us. An office worker at their desk tries to concentrate on their work task, but also communicates with other people and so is aware of the surroundings peripheral to their main activity. Dissonance in any of these pathways can produce dissatisfaction, whereas concordance and harmony produce satisfaction and even pleasure. Various senses can compensate one another, so that dissatisfaction from one maybe compensated for by satisfaction in others. Stimuli received by the eyes, ears, skin, nose and tongue are all transposed into neuron impulses, which are transmitted to appropriate parts of the brain. A new experience can be transferred from the short-term to the long-term memory if learning occurs. Repeated or similar experiences are compared with previous ones and responses are repeated or modified depending on the differences between the before and present situations, as well as the person's well-being.

The technical process is one thing, but matters such as *interpretation, psychological perception* and *imagination* are even more complex. The role of emotions and how these interact with the physiological sensory perception has fascinated philosophers and psychologists for centuries. It is often referred to as the mind–body problem. Equally, questions concerning the impacts of genetics and environment have been hotly disputed. These arguments will not be extended here, but rather it will be accepted that genetics and environment are important. The former perhaps defines the sort of person we are in terms of environmental susceptibility, but the richness and variety of environmental experience defines our development.

The environment around us can either help us to concentrate or be distracted. When we are driving a car we need to have focused concentration not only on the mechanisms of driving and seeing what is in front of us, but also having an acute awareness of peripheral stimuli from drivers behind us, traffic signs and general conditions on the road. In contrast when we are concentrating on a particular task such as writing, designing or composing the peripheral stimuli need to only provide an environmental setting, but the mind is fully focused on the particular task in hand.

Concentration requires quiet and withdrawal. Communication requires conversation and extensive interaction. Attention is the psychic energy that makes events occur in the consciousness and Heerwagen *et al.* (2005) describe evidence that states that understanding and managing attention is the single most important determinant of business success. *Attentional fatigue* has been associated with reduced performance on tasks. Attentional shifts if frequent can reduce work effectiveness. The impact of distractions on work performance is moderated by the task complexity. Distractions are more of a problem for introverts than for extroverts as well as for those carrying out highly focused creative work. For a detailed discourse on attention see Pashler (1999).

People in their workplace frequently switch attention from one thing to another; they are also distracted by other people and communications via email and phone calls. Interruptions of complex work (high information input tasks) require a longer period of adjustment. Continuous interruptions are likely to have negative effects on mood that reduce the motivation to resume the work. For simple tasks, such as repetitive work, the interruptions appear to have much less impact. In contrast however it is desirable for people to have breaks from computer work or email management; these can be considered as positive distractions that give mental breaks. Heerwagen *et al.* (2005) talk about *cognitive rest* when one may look through the window at distant views or take relief away from the workplace.

Built Environment

At the heart of architecture is the fundamental question of how buildings in their design and use can confront the questions of human existence in space-time and thus express and relate to humans being in the world. If this question is ignored the result is soulless architecture that is a disservice to humanity. There is a danger, for example, that the ever-increasing pace of technology is distorting natural sociological change and this makes it difficult for modern architecture to be coherent in human terms.

Buildings must relate to the language and wisdom of the body. If they do not they become isolated in the cool and distant realm of vision. For example, people passing by a building gain a visual impression that they like, dislike or have no particular feeling about. Buildings are a vital part of a nation's heritage and so they are historically important. This is in stark contrast to a sculptor, whose work can be selected and located according to individual choice. But in assessing the value of a building, how much attention is paid to the quality of the environment inside the building and its effects on the occupants? The qualities of the environment affect human performance inside a building and these should always be given a high priority. This can be considered as an *invisible aesthetic*, which together with the visual impact makes up a *total aesthetic* or total sensory satisfaction.

Dr Richard Jackson, state public health officer for the California Department of Health Services speaking at the University of Washington, 21 April 2005, said: "We affect the environment by what we build, but what we build, in turn, affects us and our health".

We spend most of our time in buildings and the remaining time is spent travelling between buildings, or outside experiencing the sea, countryside, practising sport or doing outdoor jobs. Inside buildings we work and live most of our lives. There are links between external and internal environments with regard to heat, light and sound. This is expressed by a preference for a view out; the need to feel connected in space and time with the outside world; the need to keep the inside–outside temperature difference to a level that will not induce thermal shock. In a sense you feel the space. A response to a particular space is a combination of the stimuli you receive via the senses, plus your particular sense of well-being at a particular moment.

Too often buildings are seen as costly static containers rather than an investment, which, if they are healthy and sustainable, can add value (see Spencer and Winch 2002; Macmillan 2006). Boyden (1971) distinguishes between needs for *survival* and those for *well-being*. Human beings have physiological, psychological and social needs. From Boyden (1971), Heerwagen (1998) pinpoints those well-being needs relevant to buildings designs as:

- social milieu;
- freedom for solitary or group working;
- opportunities to develop self-expression;
- an interesting visual scene;
- acceptable acoustic conditions;
- contrast and random changes for the senses to react to;
- opportunities to exercise or switch over from work to other stimulating activities.

To which one may add the need for clean fresh air. Stokols (1992) believes that physical, emotional and social conditions together are a requisite for good health. Buildings have a dynamic interaction with people.

One of the most important elements in considering the relationship between the indoor environment, the micro-climate around a building and the macro-climate is the building façade. We see façades; they are a very important part of our visual landscape. Behind the façade people live and work, and for them the façade provides protection from the weather (solar radiation, temperature, snow, wind and rain). The façade has a more subtle use, however, in that it is a climate modifier that can be used to control the amount of noise, sunlight and air that enters the buildings to sustain a healthy and pleasant environment. Façades combine materials like stone, brick, concrete and glass. The importance of understanding the façade has grown in recent years and façade engineering is now an accepted discipline. Smart materials and developments in biomimetics together with embedded sensor technology will revolutionise façade design. Beaven and Vincent (2004) believe natural systems are the paradigm both for smart systems and for robotics.

A primary link between the inside and outside environments is the evolutionary tie between people and nature – *biophilia* – a term coined by the Harvard social biologist E.O. Wilson (Wilson 1984; Kellert and Wilson 1993). He studied the innate affinity of humans with the natural world. Windows in the façade give this link via daylight, fresh air and views out. There are also the ephemeral stimuli originating from changing patterns of weather and time (Heerwagen 1998) giving contact with the world outside. In general people prefer natural to built-up environments; certainly landscaping of built environments is valued by people (Ulrich 1993; Kaplan and Kaplan 1989). Parsons (1991) also offers evidence that the positive effects of Nature may extend to the immune system.

Another interesting aspect is the innate affinity humans have with the underlying fractal patterns in nature even at microscopic levels.

Heerwagen (1998) calls on evidence of Humphrey (1980), Platt (1961) and Scott (1999), which suggests that fractal patterns may be a major factor in our emotional response to environments and may even contribute towards higher cognitive functioning. It seems the human sensory system welcomes variety, some degree of randomness that we will refer to as *contrast pattern recognition*, or, in Heerwagen's words, *sensory variability*. Unchanging environments are boring. Humans experience physiological and emotional daily cycles. We live through our senses and need changing patterns of stimuli to activate them (Humphrey 1980; Platt 1961; Cooper 1968; Schooler 1984; Cabanac 2005).

Architecture and the Senses

The idea of taking into account the senses of a building occupant has led to our research into how we smell, touch, hear and see things in a building, as well as our psychological interactions with them. Architecture deals not only with materials and form but also with people, emotion, space and relationships between them (Farshchi and Fisher 2000). Buildings should be a multi-sensory experience. Pallasmaa (1996, 2005) elegantly describes this belief in his book *The Eyes of the Skin* and also, in association with Holl, Pallasmaa and Perez-Gomez (1994) in the book *Questions of Perception* as does Scuri (1995) in her book on *Design of Enclosed Spaces.*

During the Renaissance, the five senses were understood to form a hierarchical system, from the highest sense of vision down to touch. This reflects the image of the cosmos in which vision is correlated with fire and light, hearing with air, smell with vapour, taste with water and touch with earth. It is by vision and hearing that we acquire most of our information from the world around us. But one should not underestimate the importance of the other senses. Olfactory enjoyment of a meal or of the fragrance of flowers and responses to temperature provide a bank of sensory experience that helps mould our attitudes towards and our expectations of the environment. The senses not only mediate information for the judgement of the intellect but they are also channels that ignite the imagination. This aspect of thought and experience through the senses is stimulated not only by the environment and people around us but, when we are inside a building, by the architecture of the space, which sculpts the outline of our reactions. Merleau-Ponty (1964) said that the task of architecture was to make visible how the world touches us.

The built environment should provide the triggers that stimulate the senses, which is the foundation of sensory experience. The question of how we select perceptions and actions is discussed by Rees and Frith (1998) and also by several authors in Salvendy (1997).

The building brief should specify the human sensory needs. However, qualitative attributes in building design briefs are often considered only at a superficial level. For example, in the case of light the level of illuminance, the glare index and the daylight factor are normally taken into account. Colour is said to be too subjective and so is often not referred to. In great spaces of architecture there is a constant, deep breathing of shadow and light; shadow inhales, whereas illumination exhales light. The light in Le Corbusier's Chapel at Ronchamps, for example, gives an atmosphere of sanctity and peace. How should we consider hue, saturation (or value) and chroma in lighting design, for example (Scuri 1995)? Goethe found Newton's scientific explanation of the nature of light limited and preferred to offer a more romantic view based on his experience of light in different situations (Bortoft 1996). In the words of Baillie (2003):

> I admire him (Goethe) for expressing the idea that we could completely lose our souls through the reductionist Newtonian approach. I think about the beautiful redness of red and how wonderful it is, but when you turn it into a number it just loses all its beauty. And that's what we do – we castrate beautiful notions turning them into numbers and theories, and we lose the soul.

We need to combine the virtues of rationalism and experience; this is what a client for a building ideally requires.

Buildings provide contrasts between interiors and exteriors. The link between them is provided by windows. Scuri (1995) considers the design of windowless environments. Beyond special situations (e.g. cinemas, studios, discos) people welcome daylight. The need for windows is complex. It includes the need for natural light, an interesting view, and

Figure 9.3 Design for the body and muscles. Animal Architecture – Exhibition, Helsinki, 1995. The floor of the exhibition space is covered by ten centimetres of sand. Source: Juhani Pallasmaa, Photo: Rauno Träskelin.

for contact with the outside world; at a fundamental level, it provides contrast for people working in buildings. It also gives one a sense of time. Much work today is done with computers at close quarters and requires the eye muscles to be constrained to provide the appropriate focal length, whereas when one looks outside towards the horizon, the eyes are focused on infinity and the muscles are relaxed. There are all kinds of other subtleties, such as the need to recreate the wavelength profile of natural light in artificial light sources, which need to be taken into account. Light affects mood. How can this be taken into account in design? Compare for example walking in a wood of trees offering dappled light with the light of an open landscape. The contrast of light and shade affects one's mood. At a practical level the *CIBSE 2012 SLL Code for Lighting* describes the importance of daylight in industry, hospitals and offices, where its effect on health and productivity have been shown to be very important (Veitch 2005).

The surfaces of buildings set the boundaries for *sound*. How a building sounds is just as important as how it looks (Shields 2003). The shape of interior spaces and the texture of surfaces determine the pattern of sound rays throughout the space. Every building has its characteristic sound of intimacy or monumentality, invitation or rejection, hospitality or hostility. A space is conceived and appreciated through its echo as much as through its visual shape, but the acoustic concept usually remains an unconscious background experience. Libeskind (2002) believes a good building is like frozen music; the walls of buildings are alive. In his words: "Buildings provide spaces for living, but are also de facto instruments, giving shape to the sound of the world. Music and architecture are related not only by metaphor, but also through concrete space."

It is said that buildings are the architecture of space, whereas as music represents the architecture of time. The sense of sound in buildings combines the threads of these notions. Without people and machines, buildings are silent. Buildings can provide sanctuary or peace and isolate people from a noisy, fast-moving world. The ever-increasing pace of change can be temporarily slowed down by the atmosphere created in a building. Architecture emancipates us from the embrace of the present and allows us to experience the slow, healing flow of time. Again, buildings provide a contrast between the passing of history and the pace of life today. The opposite is true when working with computers or watching television, for example.

One's experiences gained via the senses can evoke memories. I first heard the music of Bruckner in the newly opened Coventry Cathedral in 1962. Now a visit to this cathedral or a hearing of Bruckner's ninth symphony stirs my memory of those precious moments years ago. The combination of *feeling* the architectural space *and* the sound of music within it were powerful. At midnight on 31 December 1999, the sound of church bells rang out throughout the land. The sound was a powerful uniting experience for the nation. It made us aware of our citizenship and awakened any patriotic feelings that were within us. Recall what you feel when hearing an organ in a cathedral, or the burst of applause at the end of a concert, or the cries of seagulls in a harbour.

The most persistent memory of any space is often its *odour* (Engen 1991). "Walking through the gardens of memory, I discover that my recollections are associated with the senses", wrote the Chilean writer Isabel Allende (1998: 10). Every building has its individual scent. Think of the varying olfactory experiences such as in a leather shop, a cheese stall, an Indian restaurant, a cosmetic department or a flower shop; all awaken our memories and give, or do not give, pleasure. Wine and whisky connoisseurs know that flavour is best sensed using the nose, whereas texture is sensed in the mouth. Our sense of smell is acute, and strong emotional and past experiences are awakened by the olfactory sense. Kohler (2002) describes how Marcel Proust's childhood memories were awakened by the smell and taste of a cake recalling his Sundays with his aunt who gave him a cake dipped in her tea (Proust 1929). Odours can also influence cognitive processes that affect creative task performance as well as personal memories. Creative task performance is influenced by moods and these can be affected by odours (Warren and Warrenburg 1993; Erlichman and Bastone 1991; Baron 1990; Halbwachs 1980; Freeman 1994).

Various parts of the human body are particularly sensitive to *touch*. The hands are not normally clothed and act as our touch sensors. The skin reads the texture, weight, density and temperature of our surroundings. There is a subtle transference between tactile, taste and temperature experiences. Vision can be transferred to taste or temperature senses; certain odours, for example, may evoke oral or temperature sensations. The remarkable, world-famous percussionist Evelyn Glennie is deaf but senses sound through her hands and feet and other parts of her body (Glennie 1990). Architectural experience brings the world into intimate contact with the body. Figure 9.4 shows three columns in a Helsinki street with different shapes and made of different materials that invite people to touch and feel the temperatures and textures of each one.

The body knows and remembers. The essential knowledge and skill of the ancient hunter, fisherman and farmer, for example, can be learnt at any particular time; but more importantly, the embodied traditions of these trades have been stored in the muscular and tactile senses. Natives and their descendants from the most westerly Scottish Isle of St Kilda developed prehensile hands and feet for climbing cliff faces unaided. Architecture has to respond to behaviour that has been passed down through the genes. Sensations of comfort, protection and home are rooted in the primordial experiences of countless generations. The word *habit* is too casual and neglects the history embedded in us.

The interaction between humans and buildings is more complex than we imagine. As well as simple reactions that we can measure, there

Figure 9.4 Design for the touch. Colonnade of three shapes – three materials, Pedestrian Passage, Helsinki, 1991. Source: Juhani Pallasmaa, Photo: Rauno Träskelin.

are many sensory and psychological reactions that are very difficult to understand and quantify.

The environment inside buildings is linked to that outside by entrances (or exits), windows and chimneys. The environment inside the building has many facets. There is the *support system* provided by the facilities and various conveniences, besides information and communication systems set up to aid the links between the organisation inside and outside of the building. There is the *social environment* provided by the people themselves and there is the *physical environment* provided by the things we see, hear, touch, feel and smell. The human senses are extraordinarily sensitive and it is through them that we experience life wherever we are. It seems sensible therefore to design the environment in buildings so that the senses are fulfilled and satisfied.

The surroundings provide a rich context for the inputs to the human sensory system; *productivity* is concerned with the *outputs* from this system. At a very basic level productivity is often described as being about the speed and accuracy of carrying out particular tasks but this is a very basic description. *Quality* of working or living is much more subtle than this. Even when we are sleeping, although the sensory thresholds are raised, the environment remains important. The bed we sleep in may be uncomfortable because the body cannot rest properly. There can be interference with our sleep patterns that may not cause us to wake up but are still troublesome due to noise or excessive heat, for example. The *quality* of our sleep may be undermined by the environment.

Likewise a musician playing in a concert hall can have their performance hindered or enhanced by the environment. Of course, talented musicians will sound good whether they are playing on poor instruments, or if they are playing in concert halls with poor acoustics. Their work, and our perception of it, however, is greatly enhanced if they play on good instruments in concert halls with good acoustics. Again, the quality of their performance is influenced by the environment in which they play. People know when work is hard and difficult, and they also know when their work performance is good. *Quality work* is fulfilling work and so it is important that the environment is designed to increase the probability of achieving this. There is a double benefit here in that the individual feels more fulfilled and of course the organisation that they work for gains too.

Barriers to effective work performance are distraction, boredom, poor support systems, unhelpful organisations, a lack of social ambience and a poor physical environment. In Sweden years ago the car factories used different work shift patterns to maintain the interest of their workers. The repetitive nature of some office work and the very stressful conditions in call centres require a sensitive organisation and building for their workers. Currently television shows many interior design programmes and this no doubt affects the desires of people to make their home environment as interesting and delightful as possible. There is of course an overlap between where we live and where we work, which for some people is the same place.

Davidson (2003) led a research study at University of Wisconsin, Madison that shows positive thinking (good moods, optimism) can promote good health because the body's defences (the immune system) are stronger. This suggests that the balance between the mind and the body is a sensitive one. So how relevant is this in the workplace? Various stressors can arise from conflicts in the physical, social organisational environments. People adapt to these in various ways but for some people they will feel weakened if conditions are very stressful.

There is substantial evidence described by Heerwagen (1998) showing that positive moods are associated with the physical environment and everyday events such as social interactions (Clark and Watson 1988). Even more telling is the research that shows that positive moods aid complex cognitive strategies (Isen 1990), whereas negative moods due to distractions, discomforts, health risks or irritants arising from the physical or social environments restrict attention and hence negatively affect work performance. Because positive moods directly affect the brain processes (Le Doux 1996), it can be concluded that many aspects of building environmental design can aid task performance. Heerwagen (1998) distinguishes between direct effects such as overheating, noise or glare and indirect effects arising from mood and/or motivational factors. Several positive mood-inducing factors have already been mentioned – aesthetics, freshness, daylight, view, colour, personal control, spatial aspects and nature.

Mood, feelings and emotions affect people's decision making. Mood can be influenced by several environmental factors such as the Monday effect or weather conditions. A body of psychological literature shows that temperature is one of the important meteorological variables affecting people's mood, and this in turn influences behaviour. Cao and Wei (2005) state that research to date has revealed that stock market returns are associated with nature-related variables such as the amount of sunshine, daylight-savings time change, the length of the night and the lunar phases of the moon.

Cao and Wei (2005) describe evidence that suggests that low temperatures tend to cause aggression, and high temperatures tend to cause aggression, hysteria and apathy. The question is then "does temperature cause investors to alter their investment behaviour?" They hypothesised that lower temperature leads to higher stock returns due to investors' aggressive risk taking, and higher temperatures can lead to higher or lower stock returns since aggression and apathy have competing effects on risk taking.

Saari and Aalto (2006) quote international research that suggests that labour productivity can be increased by 2 to 8 per cent as a result of improving the indoor environment. The investments in high-quality environmental systems are recouped in a short time, because of the additional value added by the increased staff performance. With regard to temperature Seppanen et al. (2003) showed that when indoor temperatures rise above 25°C the productivity decreases by 2 per cent per degree centigrade (Seppanen et al. 2003).

Heat, light, sound, space and ergonomics are all important in designing the workplace. However, in the depth of winter or the height of summer temperature tends to be the issue that workers comment about more frequently. There is also a point that in the current sustainability agenda energy features as a highly important issue and this of course is closely related to the temperature at which we control in our buildings.

From a survey carried out by Office Angels and Union of Shop, Distributive and Allied Workers (USDAW) (see the Work section of the *Guardian*, 8 July 2006) the following conclusions were made:

- heat exhaustion begins at about 25°C;
- 24°C is the maximum air temperature recommended by the World Health Organisation for workers' comfort, but this may be higher in very hot climates to reduce thermal shock, which can occur when excessive thermal differentials are experienced between external and internal temperatures;
- 16°C is the minimum temperature recommended by the UK Work Place Regulations of 1992 (13°C for strenuous physical work);
- 78 per cent of workers say their working environments harm their creativity and ability to get the job done;
- 15 per cent of workers have arguments over how hot, or how cold, the temperature should be;
- 81 per cent of workers find it difficult to concentrate if the office temperature is higher than the norm;
- 62 per cent of workers reckon that when they are too hot they take up to 25 per cent longer than usual to complete a task.

It is worth remembering that people can die in very cold or very hot conditions. In more northern latitudes climate change has now brought about as many complaints in the summer as the winter as summer temperatures increase. Clothing is important and protocols for this vary across organisations, but some consideration needs to be given to the much hotter summers that are being experienced. One cannot often do much about one's metabolic rate in the workplace. There is the option to have natural ventilation, mechanical ventilation or air conditioning, but the temperature implications of these systems needs explaining to clients. The above survey assessed that 35 per cent of offices in the UK do not have air conditioning and rely on natural ventilation or fans. Air conditioning of course does have disadvantages in that the energy consumption is much higher than for natural ventilation systems, and also there is a slightly higher risk of building sickness syndrome.

Spatial Considerations

Space has the power to condition behaviour and form personality. . . . The environment affects our emotions, feelings and reactions.

(Scuri 1995)

The nature of space influences the way an organisation works and also how individuals feel and communicate. We are all familiar with the need for public, semi-public, group and individual specifications for space determined by the type of work being undertaken, the various communication needs and the degree of privacy required. Space also makes an impact on the building systems and the environment they create. Space affects air distribution, acoustic quality, natural ventilation and the amount of daylight. Whether a building is narrow, deep or of atrium form influences the choice of ventilation, the air conditioning and lighting systems. Again it is not only the volume that is important but also the height and the shape of the spaces together with the texture of surfaces (see Figure 9.5).

Cairns *et al.* (2001) discuss how space needs to be seen as a rich organisational resource. Space is part of the building topology and has a semiotical dimension, but it also evokes the senses, the thoughts and

Figure 9.5 Spatial expression through colour and form (note the soft and fabric-like feel of the concrete structure) in the Kuwait National Assembly, Kuwait City, 1972–1982 by Jørn Utzon.

the memories of the occupants (Gagliardi 1992). The scale, the surface textures, the light content, the sound quality, the furnishings, the colours all play their part in characterising the space in the mind of its occupants. There is an accumulated body of evidence showing that space can influence behaviour and in particular employee productivity (Bitner 1992; Aronoff and Kaplan 1995; Leaman and Bordass 1999; Clements-Croome 2000). The interplay between the senses, thoughts and memories is complex but people do react to spaces in many ways and it seems likely that a bundle of impressions and experiences are stimulated by aesthetic, functional and emotional properties of the space. Cairns *et al.* (2001) believe workplace is a psycho-physiological construct of individuals, each of whom bring their past experiences of other spaces to the present one.

Baldry (1999) points out that the relationship between the built environment and the labour process is not only a functional one but the building influences behaviour through the messages it sends out. These non-verbal communications and also the semiotics (the science of symbols and signs) together with the symbolism offered by the built environment influence behaviour (Eco 1980). Rapoport (1982) states that the environment provides cues for behaviour, whilst social situations structure behaviour. Environmental cues reinforce what is socially defined as being appropriate or inappropriate. In other words the design of buildings is partly about encoding information that users then decode (Baldry 1999). Thus the way space is structured contains and identifies symbolic messages.

Over the years there have been various patterns of space arrangements in offices from the military style hive of the 1920s and 1930s and later the open-plan office and Burolandschaft style in 1960s and 1970s. Often people ask if it is better to have open-plan or cellular offices. The answer is you need a mixture of open and cellular spaces according to the needs of the organisation. The arrangement of different kinds of spaces throughout a building creates the communication flow of the organisation via face-to-face contact and relaxation areas. Nowadays there is much more team working and it has been recognised that people need to have variations in their working day so that spaces need to be arranged to allow communications to happen at various levels. As a consequence the provision of space has become much more flexible and much more fluid, offering people a variety of settings for them to carry out their work.

A good working environment helps provide users with a good sense of well-being, inspiration and comfort. The main advantages of good environments are reduced investment in upgrading facilities, reduced sickness absence, an optimum level of productivity and improved comfort levels. Individuals respond very differently to their environments, and research suggests a correlation between worker productivity, well-being, environmental, social and organisational factors. Research shows occupants who report a high level of dissatisfaction about their job are usually the people who suffer more work and office environment related illnesses that affect their well-being, but not always so. Well-being expresses overall satisfaction. There is a connection between dissatisfied staff and low productivity; and a good sense of well-being is very important as it can lead to substantial productivity gain (Clements-Croome 2000). If the environment is particularly poor, people will be dissatisfied irrespective of job satisfaction.

Sick Building Syndrome (SBS) (Abdul-Wahab 2011)

Health is the outcome of a complex interaction between the physiological, psychological, personal and organisational resources available to individuals and the stress placed upon them by their physical and social environments, work and home life. A deficiency in any area increases stress and decreases human performance. Weiss (1997) suggests that the mind can affect the immune system. Stress can decrease the body's defences and increase the likelihood of illness, resulting in a lowering of well-being. Stress arises from a variety of sources: *the organisation, the job, the person* and *the physical environmental conditions*. It can affect the mind and body, weaken the immune system and leave the body more vulnerable to environmental conditions. In biological terms, the hypothalamus reacts to stress by releasing the hormone ACTH; then the hormone cortisol in the blood increases to a damaging level possibly affecting the brain cells involved in memory. This chain of events interferes with human performance, and productivity falls as a consequence.

Many surveys have shown that people can feel unwell whilst they work in buildings but recover when they leave them. The SBS symptoms usually are rooted in respiratory, eyes, cerebral, skin or musculoskeletal discomforts, which may exhibit themselves as minor irritations or even pain. Cerebral conditions include headaches and unusual tiredness or lethargy.

Some people are more susceptible than others. There can also be a social chain reaction. Nevertheless numerous independent surveys cannot all be wrong. The causes can be a combination of various factors including lighting, ventilation, contaminants in the air, changes in and levels of temperature, relative humidity or acoustic conditions. The immune system can be lowered by stresses around us that may be social, organisational, personal or environmental. Once lowered by one factor the body's defence system becomes more vulnerable to other factors.

An underlying hypothesis is that sick building syndrome is caused by building-related factors. Berglund and Gunnarsson (2000) question this postulate and ask if there is a relationship between the personality of the occupant and sick building syndrome. Certainly some people complain about various issues more than others; some people are much more sensitive and therefore much more susceptible to environmental influences than others. They conclude that personality variables can account for about 17 per cent of the SBS variants. In their work on the effect of environment on productivity Clements-Croome and Li (2000) propose a holistic model that considers the impact of the social ambience, organisation, well-being of the individual and physical environment factors and derive relationships between productivity and job satisfaction, stress, physical environment, SBS and other factors. This multifunctional approach is applicable in real life situations.

We live through our senses and the environment we provide for them to interact with is important. A building and its environment can help people produce better work, because they are happier and more satisfied when their minds are concentrated on the job at hand; building design can help achieve this. At low and high arousal or alertness levels, the capacity for performing work is low; at the optimum level the individual can concentrate on work while being aware of peripheral stimuli from the physical environment. Different work requires different environmental settings to achieve an optimum level of arousal. It is necessary to assess if a sharper or leaner indoor environment is required for the occupants' good health and high productivity and to redefine comfort in terms of *well-being*.

People spend about 90 per cent of their lives in buildings, so the internal environment has to be designed to limit the possibilities of infectious disease; allergies and asthma; and building-related health symptoms, referred to as sick building syndrome symptoms. Buildings should provide a multi-sensory experience, and therefore anything in the environment that blocks or disturbs the sensory systems in an unsatisfactory way will affect health and work performance. Thus, lighting, sound, air quality and thermal climate are all conditions around us that affect our overall perception of the environment. Air quality is a major issue because it only takes about four seconds for air to be inhaled and for its effect to be transmitted to the bloodstream and hence the brain. Clean, fresh air is vital for clear thinking, but it is not the only issue to be considered.

Fisk (1999) discusses linkages between infectious disease transmission, respiratory illnesses, allergies and asthma, sick building syndrome symptoms, thermal environment, lighting and odours. He concludes that in the US the total annual cost of respiratory infections is about $70 billion, for allergies and asthma $15 billion, and reckons that a 20–50 per cent reduction in sick building syndrome symptoms corresponds to an annual productivity increase of $15–38 billion. The linkage between odour and scents and work performance is less understood, but Fisk (1999) concludes that the literature provides substantial evidence that some odours can affect some aspects of cognitive performance. He refers to work by Rotton (1983), Dember *et al.* (1995), Knasko (1993), Baron (1990) and Ludvigson and Rottman (1989). The application of scents has been used by the Kajima Corporation in their Tokyo office building, as reported by Takenoya in Clements-Croome (2000).

Fisk goes on to consider the direct linkage between human performance and environmental conditions and writes that for US office workers there is a potential annual productivity gain of $20–200bn. His conclusions are that there is relatively strong evidence that characteristics of buildings and indoor environments significantly influence the occurrence of respiratory disease, allergy and asthma symptoms, sick building syndrome and worker performance. In 2002 the total sick leave due to stress-related illnesses cost the UK £376 million; a significant part of this was due to the physical environment.

Roelofsen (2001) describes a study of 61 offices (7,000 respondents) in the Netherlands that showed people were away for 2.5 days per year on average because of unsatisfactory indoor environmental conditions. This represented a quarter of the total average absenteeism. Other work by Preller *et al.* (1990) and Bergs (2002) reveals a close correlation between sick leave and building-related health complaints.

Conclusions

The environment matters in all the ways that have been described. It is an intrinsic part of our existence. Here are some quotes from the Royal Society for the Promotion of Health Annual Lecture delivered in Scarborough on 7 June 2005 by John Sorrell the Chairman of Commission for Architecture and the Built Environment (CABE):

> We know that good design provides a host of benefits. The best designed schools encourage children to learn. The best designed hospitals help patients recover their health. Well-designed parks and town centres help to bring communities together . . .

> But true delight goes beyond the issue of beauty, it must also consider how the building contributes to the experience of those who use it, and whether it also makes a positive contribution to the community in which it is based . . .

> The design of our work places can also have a fundamental impact on occupational health. 14 million days are lost each year in the UK through absenteeism from which at least 70 per cent of which is related to health issues . . .

> In our schools classrooms with good daylight, natural ventilation and good acoustics have been shown to have a significant impact on educational achievement . . .

> And yet when, in 2004, CABE asked members of the public to comment on their experience of hospital environments, 83 per cent of the comments were negative. These are some of the expressions used to describe that experience:

> Cold, depressing, dehumanisisng, Kafka-esque, dirty, smelly, frightening, impersonal, confusing, dull, shabby, windowless, grim, over-crowded, Gormenghast, no personality, stressful. Unpleasant, little natural light or air, harsh, disorientating, designed to confuse, no privacy.

The environment plays a vital part in our personal and working lives. Design needs to recognise this and clients need to be shown that high-quality design is an investment that increases business value. Environment matters.

References

Abdul-Wahab, S.A., 2011, *Sick Building Syndrome* (Berlin: Springer).

Allende, I., 1998, *Aphrodite: a Memoir of the Senses* (London: Flamingo).

Aronoff, S. and Kaplan, A., 1995, *Total Workplace Performance: Rethinking the Office Environment* (Ottawa: WDL Publications).

Baillie, C., 2003, Science with Soul, *RSA Journal*, April, 38–41.

Baldry, C., 1999, Space: The Final Frontier, *Sociology*, 33, 3, 1–29.

Baron, R.A., 1990, Environmentally Induced Positive Effect: Its Impacts on Self-efficacy, Task Performance, Negotiation and Conflict, *Journal of Applied Social Sociology*, 20, 5, 368–384.

Beaven, M. and Vincent, J., 2004, Engineering Intelligence through Nature, in D. J. Clements-Croome (ed.) *Intelligent Buildings* (London: Thomas Telford).

Berglund, B. and Gunnarsson, A.G., 2000, Relationships between Occupant Personality and the Sick Building Syndrome Explored, *Indoor Air*, 10, 152–169.

Bergs, J., 2002, The Effect of Healthy Workplaces on the Well-Being and Productivity of Office Workers, Plants for People Symposium, Reducing Health Complaints at Work, 14 June, Amsterdam.

Bitner, M., 1992, Servicescapes: the impact of physical surroundings on customers and employees, *Journal of Marketing*, 56, 2, 37–72.

Bortoft, H., 1996, *The Wholeness of Nature: Goethe's Way of Science* (Edinburgh: Floris Books).

Boyden, S., 1971, Biological Determinants of Optimal Health, in D.J.M. Vorster (ed.) *The Human Biology of Environmental Change*. Proceedings of a conference held in Blantyre, Malawi, April 5–12 (London: International Biology Programme).

Cabanac, M., 2005, Pleasure and joy, and their role in human life, in D.J. Clements-Croome (ed.) *Creating the Productive Workplace* (London: Routledge).

Cairns, G., McInnes, P. and Roberts, P., 2001, Organizational space/time: From imperfect panoptical to heterotopian understanding, *Ephemera: Critical Dialogues on Organization*, 3, 2, 1473–2866.

Cao, M. and Wei, J., 2005, Stock market returns: A note on temperature anomaly, *Journal of Banking & Finance*, 29, 1559–1573, www.yorku.ca/mcao/cao_wei_JBF.pdf (accessed 1 July 2005).

Clark, L.A. and Watson, D., 1988, Mood and the Mundane: Relationships Between Daily Events and Self-Reported Mood, *Journal of Personality and Social Psychology*, 54, 296–308.

Clements-Croome, D.J. (ed.), 2000, *Creating the Productive Workplace*, second edition (London: Spon-Routledge).

Clements-Croome, D.J. and Li, B, 2000, Productivity and Indoor Environment, Proceeding of Healthy Buildings Conference, Helsinki, University of Technology, 6–10 August, 1, 629–634.

Cooper, R., 1968, The Psychology of Boredom, *Science Journal*, 4, 2, 38–42.

Davidson, R.J., 2003, Report by M. Henderson, *The Times*, 2 September, 4.

Dember, W.N., Warm, J.S. and Parasuraman, R., 1995, Olfactory stimulation and sustained attention, in Gilber, A.N. (ed.) *Compendium of Olfactory Research: explorations in aroma-*

chology: investigating the sense of smell and human response to odours. 1982–1994 (Iowa: Kendall Hunt Pub. Co.).

Eco, U., 1980, Functions and Signs: The Semiotics of Architecture, in Broadbent and Jencks (eds.) *Signs, Symbols and Architecture* (Chichester: John Wiley).

Engen, T., 1991, *Odour Sensation and Memory* (Westport, CT: Praeger).

Erlichman, H. and Bastone, L., 1991, Odour Experience as an Affective State, Report to the Fragrance Research Fund, New York.

Farshchi, M.A. and Fisher, N., 2000, Emotion and the environment: the forgotten dimension, in D.J. Clements-Croome (ed.) *Creating the Productive Workplace* (London: Spon).

Fisk, W.J., 1999, Estimates of potential nationwide productivity and health benefits from better indoor environments: An update, Lawrence Berkeley National Laboratory Report, in J.D. Spengler, J.M. Samet and J.F. McCarthy (eds) *Indoor Air Quality Handbook* (New York: McGraw Hill).

Freeman, W.J., 1994, Chaotic Oscillations and the Genesis of Meaning in the Cerebral Cortex, http://sulcus.berkeley.edu/wjf/AB.Genesis.of.Meaning.pdf (accessed 18 February 2013).

Gagliardi, P., 1992, *Symbols and Artifacts: View from the Corporate Landscape* (New York: Aldine de Gruyter).

Glennie, E., 1990, *Good Vibrations* (London: Hutchinson).

Greenfield, S., 2000, *Brain Story* (London: BBC Publications).

Halbwachs, M., 1980, *The Collective Memory* (New York: Harper and Row).

Heerwagen, J.H., 1998, Productivity and Well-Being: What are the Links? American Institute of Architects Conference on Highly Effective Facilities, Cincinnati, 12–14 March.

Heerwagen, J. *et al.*, 2005, The Cognitive Workplace, in D.J. Clements-Croome (ed.) *Creating the Productive Workplace* (London: Routledge).

Holl, S., Pallasmaa, J. and Perez Gomez, A., 1994, *Questions of Perception* (Special Issue by *Architecture and Urbanism*).

Humphrey, N., 1980, Natural Aesthetics, in B. Mikellides (ed.) *Architecture for People* (London: Studio Vista).

Isen, A., 1990, The Influence of Positive and Negative Affect on Cognitive Organisation: Some Implications for Development, in N.L. Stein, B. Leventhal and T. Trabasso (eds.) *Psychological and Biological Approaches to Emotion* (Hillsdale, NJ: Erlbaum).

Kaplan, S. and Kaplan, R., 1989, *The Experience of Nature: A Psychological Perspective* (Cambridge: Cambridge University Press).

Kellert, S. and Wilson, E.O., 1993, *The Biophilia Hypothesis* (Washington, DC: Island Press, Shearwater Books).

Knasko, S.C., 1993, Performance Mood and Health During Exposure to Intermittent Odours, *Archives of Environmental Health*, 48, 5, 305–308.

Kohler, N., 2002, Sustainability and Indoor Air Quality, Proceedings of 9th Int. Conference on Indoor Air Quality and Climate, Monterey, 30 June–5 July, 4, 1–9.

Leaman, A. and Bordass, B., 1999, Productivity in buildings: the "killer" variables, *Building Research and Information*, 27, 1, 4–19.

Le Doux, J., 1996, *The Emotional Brain* (New York: Simon and Schuster).

Libeskind, D., 2002, The Walls are Alive, *Guardian*, 13 July.

Ludvigson, H.W. and Rottman, T.R., 1989, Effects of odours of lavender and cloves on cognition, memory, affect, and mood, *Chemical Senses*, 14, 4, 525–536.

Macmillan, S., 2006, Added Value of Good Design, *Building Research and Information*, 34, 3, 257–271.

Merleau-Ponty, M., 1964, *Eye and Mind in Primary of Perception* (Evanston, IL: Northwestern University Press).

Pallasmaa, J., 1996, *The Eyes of the Skin: Architecture and the Senses* (London: Academy Editions).

Pallasmaa, J., 2005, *The Eyes of the Skin: Architecture and the Senses*, second edition (Chichester: Wiley-Academy).

Parsons, R., 1991, The Potential Influences of Environmental Perception on Human Health, *Journal of Environmental Psychology*, 11, 1–23.

Pashler, H.E., 1999, *The Psychology of Attention* (Cambridge, MA: MIT Press).

Platt, J.R., 1961, Beauty: Pattern and Change, in D.W. Fiske and S.R. Maddi (eds) *Functions of Varied Experience* (Homewood, IL: Dorsey Press).

Preller, L., Zweers, T., Brunekreef, B. and Boleiji, J.S.M., 1990, Indoor Air Quality '90, Fifth International Conference on Indoor Air Quality and Climate, 1, 227–230.

Proust, M., 1929, *Swann's Way* (London: Chatto and Windus).

Rapoport, A., 1982, *The Meaning of Built Environment* (London: Sage).

Rees, G. and Frith, C.D., 1998, How Do We Select Perceptions and Actions? *Philosophical Transactions, Brain Mechanisms of Selective Attention and Action*, August, 353, 13713, 1283–1293.

Roelofsen, 2001, The Design on the Workplace as a Strategy for Productivity Enhancement, Presented at the 7th REHVA World Congress, Clima 2000, Naples, 15–18 September.

Rotton, J., 1983, Affected and Cognitive Consequences of Malodorous Pollution, *Basic and Applied Psychology*, 4, 2, 171–191.

Saari, A. and Aalto, L., 2006, Indoor Environment Quality Contracts in Building Projects, *Building Research and Information*, 34, 1, 66–74.

Salvendy, G (ed.) 1997, *Handbook of Human Factors and Ergonomics* (Hoboken, NJ: Wiley Interscience).

Schooler, C., 1984, Psychological Effects of Complex Environments During the Life Span: A Review and Theory, *Intelligence* 8, 259–281.

Scott, S.C., 1999, *Visual Attributes Related to Preferences in Interior Environments*, unpublished manuscript.

Scuri, P., 1995, *Design of Enclosed Spaces* (London: Chapman Hall).

Seppanen, O., Fisk, W. J. and Faulkner, D., 2003, Cost–Benefit Analysis of the Night Time Ventilative Cooling in Office Building, in *Proceedings of Healthy Buildings*, Singapore, 3, 394–399.

Shields, B., 2003, Learning's Sound Barrier, by Nina Morgan in *Newsline*, 26, 10–11.

Spencer, N. C. and Winch, G. M., 2002, *How Buildings Add Value for Clients* (London: Construction Industry Council and Thomas Telford).

Stokols, D., 1992, Establishing and Maintaining Healthy Environments: Toward a Social Ecology of Health Promotion, *American Psychologist*, 47, 1, 6–22.

Ulrich, R. S., 1993, Biophilia, Biophobia, and Natural Landscapes, in S.K. Kellert and E. O. Wilson (eds.) *The Biophilia Hypothesis* (Washington, DC: Island Press, Shearwater Books).

Veitch, J., 2005, Creating High Quality Workplaces Using Lighting, in D.J. Clements-Croome (ed.) *Creating the Productive Workplace* (London: Routledge).

Warren, C. and Warrenburg, S., 1993, Mood Benefits of Fragrance, *Perfumer and Flavourist*, 18, March/April, 9–16.

Weiss, M.L., 1997, *Division of Behavior and Cognitive Science*, PhD thesis, Rochester University, New York.

Wilson, E.O., 1984, *Biophilia: The Human Bond with Other Species* (Cambridge, MA: Harvard University Press).

Chapter 10

Biophilia, Topophilia and Home

Boon Lay Ong

The root word of both ecology and economics is a Greek word, *οἶκος*, meaning *house*. Ecology is thus the study of our *home* while economics is the study of the *management (nomos)* of the home. Ecology is defined as a study of the relationship between living things and the natural environment (including other living things and each other). Strangely enough, ecology does not include the built environment as part of its central concern. *Urban ecology* in most texts refers to the wild animals and plants that survive in the built environment rather than the built environment itself. Indeed, often in spite of our efforts rather than because of them. Even *human ecology*, as a field of study, is concerned with our relationship with the natural environment rather than with how the built environment itself functions as a system. On the other hand, economics is not at all interested in the natural world. And, in our current state of environmental crisis, ecological concerns seem to be in conflict with economic ambitions. Is not the *house* of ecology also the *house* of economics?

E.O. Wilson (1984) defined *biophilia* as "the connections that human beings subconsciously seek with the rest of life". In recent years, there has been quite a substantial amount of research that elaborates on how biophilia is manifested. Much of this work is well summarised in two books: *The Biophilia Hypothesis* (Kellert and Wilson 1993) and *Biophilic Design: Theory, Science and Practice of Bringing Buildings to Life* (Kellert, Heerwagen and Mador 2008). Our bond with nature stems not just from the evolutionary past when we lived in direct dependence upon nature but also as a result of our continuing dependence on the providence of nature despite technological advances. Heerwagen (2009: 40) describes it thus:

> The sun provided warmth and light as well as information about
> time of day. Large trees provided shelter from the midday sun

and places to sleep at night to avoid terrestrial predators. Flowers and seasonal vegetation provided food, materials, and medicinal treatments. Rivers and watering holes provided the foundation for life – water for drinking and bathing, fish and other animal resources for food. Waterways also provided a means of navigation to reach distant lands.

The built environment we live in today may not make these dependencies explicit and the role of nature in our everyday lives may seem far removed but these provisions still exist and we still recognise nature for its value in these ways. The difference is that whilst our forefathers may have felt the presence of nature more starkly, our bonds with nature are perhaps more distant and aesthetic. The key point to note here though is that the biophilic bond is a reflection of our ecological relationship with the natural world. It is a relationship that goes back into our distant past and lies unconscious in our reading of the landscape.

Ecological Place

Esther Sternberg (Sternberg 2009) further elaborates on this with a collation of scientific evidence linking place and well-being. A medical doctor, she points out that there is a turning point in illness when healing begins. This point coincides with a sudden noticing of the environment around you "when your interest in the world revives and when despair gives way to hope" (Sternberg 2009: 1). She asks: "Can we design places so as to enhance their healing properties? And [conversely] if we ignore the qualities of physical context, could we inadvertently slow the healing process and make illness worse?" She answers affirmatively: "The idea that physical space might contribute to healing does, it turns out, have a scientific basis."

Sternberg explores this relationship between us and place not just through our various senses but also in terms of mystery (labyrinths), wayfinding, and even hope, thought and prayer. Our modern mind is taken aback by the suggestion that physical place can even affect, if not effect, our thoughts, hopes and prayers. Are not spiritual matters by definition separated from the material? It is a great miscarriage that in our pursuit of visual aesthetics to the exclusion of every other mode of perception and understanding, modern architecture is discussed and focused on appreciation only from a distance and outside of human engagement. If we do not understand architecture as ecological place, how can we hope to design architecture that is not just visually appealing but also comforting and supportive of our well-being? True, the best of even modern architecture fulfils our sense of place in all dimensions but this concern and capability is not transmitted in our literature and discussed through our discourse.

In *Topophilia* (1974), Yifu Tuan recounts the many dimensions of our relationship with the land. From primordial psychological structures of cosmological schemata to the search for new artificial utopias like the modern suburb and urban living, Tuan discusses the variety of ways in which culture and knowledge evolved from interpretations of the environment. He points to two opposing and yet complementary concepts: the city and the wilderness. While our security and livelihood is most easily found in the bustle of the city, we yearn for the wilderness in pursuit of peace and solitude. The garden (of Eden) evolved to fill the middle ground in an attempt to answer both needs. And yet, the movement towards urbanisation appears to be relentless. Currently, around half of the world's population live in cities and the proportion looks set to grow.

These different ideas are connected in architecture. The making of architecture is also a making of place, an appropriation of a certain location or site and demarcating it for human purpose. What these ideas point to is that the making of place is an ecological act that, on the surface at least, tells of the division and perhaps even antagonism between the manmade and the natural. And yet, lying beneath this surface, is a deep bond and dependency with nature that also needs recognition. This expression of our complex relationship with nature, on the one hand dependent and on the other antagonistic, may be found in many great architectures and may even be considered a key thematic concern in great architecture. E.O. Wilson (Kellert *et al.* 2008: 23) elaborates:

> . . . people want to be in the environments in which our species evolved over millions of years, that is, hidden in a copse or against a rock wall, looking out over savanna and transitional woodland, at acacias and similar dominant trees of the African environment. And why not? Is that such a strange idea? All mobile animal species have a powerful, often highly sophisticated inborn guide for habitat selection. Why not human beings?

I had the opportunity to explore some of these themes in a house I designed for a friend.

An Ecological Approach to Home

In nature, ecology is driven by cosmology. The revolution of the planets with respect to the sun, the pull of the moon and the rotation of our earth determine the overarching characteristics of our climate, the rhythms of day and night and the changing of the seasons. The answer to whether or not there will be a good harvest this summer is found not in the soil that we can till but in the planets above, over which we have no say. Historically, the layout and design of architecture and cities are bound with cosmological ideologies. The modern mind may see these links as mere superstition having no basis in reality nor consequence. To the ancient mind, however, the cosmos is the larger nature that rules ecology, the closer nature within which we dwell.

In approaching the design of a house for a friend, I was concerned with the idea of place making. This in itself is no great revelation as all good architecture should be good places as well. As the design developed, I became aware of the ecological metaphors that the house acquired in relation to the land upon which it sits. Here was a waterfall, there a bridge, a cave and tunnel, steps like branches spiralling up the trunk into the shade of a canopy and views out into the distance. I realised then that these are archetypal forms that are found in nature and resonate in us even when encountered in the built environment. The house isn't just an ecosystem, it is an aggregation of ecosystems, albeit perhaps more metaphorically than biologically.

Waterfall

One of the most powerful and poignant natural landforms is the waterfall. Often a sacred site, the waterfall beckons with its gentle refrain, like the soothing hushing of a mother. Up close, natural waterfalls are often loud enough to be threatening and a conscious effort has to be made to get up closer. Big or small, they are always a natural attraction capable of drawing people from quite a distance away to come and enjoy the water. They are quite the most persistent of all natural places.

Perhaps the most famous image of a house of the twentieth century is the waterfall view of Frank Lloyd Wright's Fallingwater. Here, you do not actually get a view of the waterfall unless a deliberate trip is made to the bottom of the falls and you look back. The entrance to the house is from the back and the waterfall makes itself known through its sound, and hence its name. The hiding of the waterfall from view is of course deliberate and necessary – not just to emphasise the essential nature of a waterfall but also to respect its visual beauty. You respond to the call of a lover before you see her.

In my case, the perching of the house at the top of the slope meant that the waterfall is seen upon arrival at the house (Figure 10.1). Fortunately, there is a turning before the entrance and there is still an element of surprise in the encounter. Also, the modern practice of coming home in an enclosed air-conditioned car means that acoustical fancies like the pink noise of a waterfall will go unnoticed until one has parked the car and gotten out.

Paradise

Above the waterfall is a suspended garden. Cantilevered out about 10 metres, the living room is designed as an indoor garden. The living room itself is set back two metres from the edge so that it may be surrounded by greenery. The two metres setback helps to provide shade while floor-to-ceiling glazing allows for a full view of the surrounding terrain. From here, the client can see people arriving at the front gate, look back at the pool, the garden courtyard, the rest of the house and then beyond

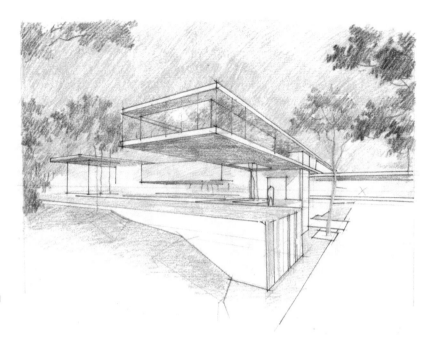

Figure 10.1 The gate is to the right and arrival at the house will receive a frontal view of the waterfall and hanging garden in the living area.

into the distance. A panoptic perch, the cantilevered living room symbolises the nest, a "coming home after a hard day's work". It is the soul of the house.

For many people, *paradise* is not just a garden but the *ideal* home. It is a place *par excellence* where we need no artificial shelter and feel no threat. Different myths of paradise exist in the imagination of all cultures and hark back to our primordial past where we once lived in total dependence on the providence of nature. In our imagination, our difficulties are forgotten and seem insignificant in the light of current worries. To be successful, our homes must be a paradise, an escape from harsh reality, the best of all dreams. Properly, then, paradise as the centre and symbol of home is here both a garden and suspended from the ground. The entrance to the house is to the right, as you turn in the driveway and stop beneath an overhanging bridge (Figure 10.2). The garage is to the back and the entrance to the house just beneath the bridge and to the left.

Cave

The cave is most likely our first home. The cave is the only natural structure that will afford us substantive protection from the elements (rain, snow, wind) and from wild animals. There is only the cave opening to defend once we have cleared the inside of all threats and unwanted material. Its darkness is our hiding place and the air, though dank, is still likely cooler

Figure 10.2 Turning to the right, you arrive at the porch over which hangs a bridge. The driveway leads to the garage at the back.

than a warm tropical night and warmer than a cold temperate winter. Though few amongst us can remember back quite so far, we think of the cave as a womb where we lie protected in its warm and wet darkness.

The entry hall to the house is in the lowest floor – part basement because it lies beneath the main level of the house and part ground floor because it lies on the same plane as the roadway that leads into the house. On this floor is the games room and a home spa, both of which open into a sunken garden. It is a meditative garden, full of flowers and other orna-mental plants (the owner has a love for bonsai plants) where one might linger to pore over a leaf or a flower. Tuan says of a Chinese garden: "To walk into a Chinese garden and be aware of even a fraction of its total meaning is to enter a world that rewards the senses, the mind, and the spirit" (1974: 146).

This lowest floor is a place of repose, of contemplation, of getting close to the earth. The materials here are dark and rough, with just enough light carefully positioned to accentuate the mystery.

Bridge

If the cave is our first home, the bridge is likely our first structure. Encountered sometimes as trees that fell across a stream or a ravine, our intelligent forefathers will have been quick to recognise the potential of building bridges to get across physical breaks in their paths. When trying to explain the phenomenology of dwelling, Heidegger used the bridge as an example. The bridge sets off the environment around it, the banks on either side, the water, the sky and even the earth and the water. The bridge

Figure 10.3 The entry hallway of rough hewn stone leading to a top-lit spiral staircase that leads to the main level of the house.

allows the impossible to happen: for the water to run its course and for people to cross the river at the same time. In Heidegger's account, the bridge "leads" by enabling transport and trade, linking the village to the world beyond, the farmer to his market. "Always and ever differently the bridge escorts the lingering and hastening ways of men to and fro, so that they may get to other banks and in the end, as mortals, to the other side" (Heidegger 1971: 152–153).

The bridge over the ravine created by excavating the existing landform provides a natural archway to welcome the homeowner – a porch or portico. According to Heidegger, "where the bridge covers the stream, it holds its flow up to the sky by taking it for a moment under the vaulted gateway and then setting it free once more".

Verticality

Both Bachelard (1969) and Tuan (1974) emphasise the vertical as a dimension in nature that is fraught with meaning. The verticality takes form in Bachelard in his oneiric house where the middle ground is made for living, the attic a place of dreams and positive memories and the basement a place of fear and negativity. These meanings of vertical space are not restricted to the house but influence the house because of the significance of earth, sky and abyss in nature. If we fall, we fall to our death, injury or some equally distressing fate. If we climb, we climb to higher ground, a better view and safety. The light is brighter as we climb up and darker as we move down.

Usually, verticality is traversed in nature by slopes but we also climb trees to reach a higher plane. Climbing a tree is perhaps the primordial equivalent of climbing a spiral staircase. In the Bible, it is a stairway that leads Jacob to heaven (Genesis 28:10–22). As in Heidegger's bridge, the stairway links (leads and enables) two otherwise separated places. Climbing a stairway, like Jack and the Beanstalk, leads from one world to a different other world.

In the house, a spiral staircase leads from the dark earthy basement/ground floor to the bright middle/main floor. It is this light that shines at the end of the tunnel of the entrance hallway.

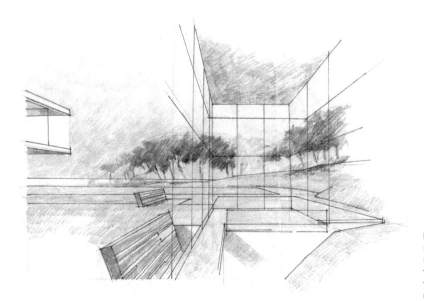

Figure 10.4 The top of the stairway is the reception space for the house – the family lounge from which the rest of the house unfolds.

Figure 10.5 The spiral staircase from the basement rises up to the top storey as the courtyard, pool and enclosing house is revealed.

Clearing

The forest is rich with possibilities but too crowded for ideal human dwelling. Invariably, we make a clearing upon which to live. Orians and Heerwagen (1992) argue that we have a preference for clearings because we evolved from the Savannah. Certainly, the open spaces of courtyards, grass lawns and flat plains are attractive not least because it is harder for threats to lie hidden. A flat plain is also better suited to our bipedalism. Tuan (1974: 79–80) provides further evidence that in the all-enveloping closeness of the forest, there are no horizons, views to distant landmark, mountain or isolated tree to distinguish an object from the rest. Consequently, the mythology of the Pygmies, for example, are little concerned with "the creation of the world, with the stars and sky". Even their perspective of time is short, concerned more with the present than the past. With open space, the skies are revealed and with it the diminutive scale of human life and reach compared to the space-time of the universe. It is in the open space that we develop our cosmological awareness and schemata.

The paradise garden is such a clearing. It encapsulates the fundamental elements of ecology – the stream, plants and animals – in a schema that points to the four directions and the four quarters of the earth. Vita Sackville-West was quoted by Brookes (1987: 13–14):

Imagine you have ridden in summer for four days across a plain; that you have then come to a barrier of snow-mountains and ridden up the pass; that from the top of the pass you have seen a second plain, with a second barrier of mountains in the distance, a hundred miles away; that you know that beyond these mountains lies yet another plain, and another; and that for days, even weeks, you must ride with no shade, and the sun overhead, and nothing but the bleached bones of dead animals strewing the track. Then when you come to trees and running water, you will call it a garden. It will not be flowers and their garishness that your eyes crave for, but a green cavern full of shadows and pools where goldfish dart and the sound of a little stream.

Arrival, Dwelling and Home

The sense of place is felt as a sense of arrival. The site and space invites you to stay. This can happen in many ways: with the hint that there is more to discover, perhaps a sense of abundance just for the taking, perhaps a vague familiarity and perhaps with a simple feeling that it is a good place to rest. The thesis forwarded here is that the sense of place in architecture is related to the kinds of places we find in nature and that these places reflect our ecological relationship with natural environments.

Figure 10.6 The central courtyard is a repose with the cantilevered living room to the right and the family lounge to the left.

In many ways, these places work through metaphors. Metaphors are generally understood today as just references or representations that are at best symbolic or poetic. In practice, though, metaphors imbue the object with the properties of the metaphoric reference. To the ancient and receptive mind, to say that a particular mountain is a dragon is not so much a self-delusional statement but attributing properties of the dragon to the mountain, and instigating the same fear for the mountain that one might have for the dragon. The power of the horoscope in many cultures relies on metaphor to translate inherent characteristics of people born under particular stars and constellations. These metaphors are not just convenient comparisons but guides to reading the constellations and interpreting their meanings. To recognise a face in the shape of the mountain does not just make the mountain visually interesting but lifts it into the realm of the supernatural, gives it magical properties and protects it as sacred landscape. Metaphor in architecture does not simply refer to another but helps create place by imparting properties of metaphorical place into architectural space.

Exercising the mind metaphorically brings greater benefits than just making a place seem more magical and meaningful. Michael Michalko (Michalko 1998) studied the minds of geniuses and found that metaphor is often used in their breakthroughs. Einstein famously did a thought experiment where he travelled along the path of a light ray, Archimedes had his eureka moment in a bathtub and Stephen Hawking drew pictures that his assistant then had to translate into mathematics. Designing for comfort alone can lull the mind and tire the brain. The right kind of stimulation provides a restfulness that is entertaining and refreshing.

The construction of the spaces described here includes the more technical concerns of ensuring enough light, providing thermal comfort and the other aspects of environmental comfort. Yet, these concerns are like parts of a recipe. Both the cook and the diner are interested in the food at the end rather than the recipe itself. Like good food, the home is not just about physical satisfaction and comfort. It is also about discovery and indulgence, taking the inhabitant into a world within that both protects and prepares him for the world without.

Hydrology and the Wisdom of God

In 1968, Yifu Tuan wrote a book with a curious title, *The Hydrologic Cycle and the Wisdom of God*. In it, he discussed the notion held by scientists in the early days of the scientific revolution, in this particular case, during the seventeenth century, that the rationality of scientific explanation reveals the existence and wisdom of the Christian God. The hydrologic cycle, described simplistically as the circulation of water from the oceans to the land and back again, is of primary significance to the ecology of the planet. A central component of climate in the form of rain, snow and humidity, it is also the key process of distributing water and nutrients to the various

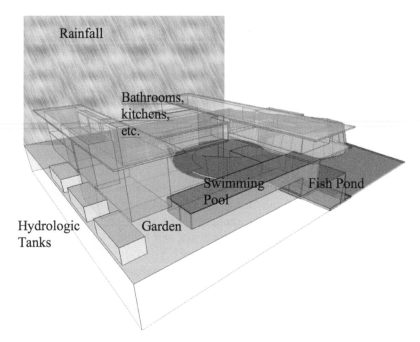

Figure 10.7 Hydrological system of the house: collection of rainwater; separation of water into clean, drinkable, grey and black; filtration and treatment using plants, biological filtration and reverse osmosis.

ecosystems of the planet, thus enabling the diversity of life itself. The question was, why did God provide so much water? There is approximately three times as much ocean as there is land. Why the abundance of water when humans dwell mainly on land?

The sceptic would immediately suggest that this proves that God is not anthropocentric but the theological argument need not concern us here. What is of greater importance is that then, as now, the hydrologic cycle lies at the centre of a sustainable ecology. In most modern buildings, clean potable water is piped in from the mains and used water of different levels of pollution is then drained away. The sustainable home cannot afford to continue this wasteful process and must incorporate as a central strategy an internal hydrological cycle to recycle the water used in the house as much as possible. Fortunately, with current technology, this is entirely possible and not particularly expensive.

The Ecological Home

Home is presented here as dwelling in nature, not as wild nature, but nature accommodated to our sensibilities. The call for sustainability is biophilic and appeals to our intrinsic natural selves as well. To look upon nature as purely a provision shop for our needs is to deny a deep appreciation for nature that is as much a part of us as any of our deepest needs. To deny that we have needs not fulfilled by nature alone is to deny some of the more laudable characteristics of humanity. Most cultures, in their literature, arts and architecture, suggest ways in which our relationship and dwelling in nature can be mutually beneficial – that we might be recharged by nature and that we might, in turn, tend to nature as if it were our garden.

The authors in this book urge in their expert wisdom and passionate voices that we need to go beyond science to understand the full consequence of the results it produces. It is apt that this requires that we embrace the poetic, the emotional and the intuitive in us – areas long denied access because of the rationality and hegemony of science. Ecology is all about balance and when our two aspects, aesthetics and science, are thus balanced and validate one another, we will finally find our home.

References

Bachelard, G. (1969) *The Poetics of Space*. Translated by Maria Jolas, foreword by Etienne Gilson. Boston, MA: Beacon Press.

Brookes, J. (1987) *Gardens of Paradise: the history and design of the great Islamic gardens*. London: Weidenfeld and Nicolson.

Heerwagen, J. (2009) Biophilia, health, and well-being. In Campbell, L. and Wiesen, A. (eds) *Restorative Commons: creating health and well-being through urban landscapes*. Gen. Tech Rep. NRS-P-39. US Department of Agriculture, Forest Service, Northern Research Station: 38–57.

Heidegger, M. (1971) Poetry, Language, Thought. Translation and introduction by Albert Hofstadter. New York: Harper & Row.

Kellert, S. R., Heerwagen, J. H. and Mador, M. L. (eds) (2008) *Biophilic Design: theory, science and practice of bringing buildings to life*. Hoboken, NJ: John Wiley.

Kellert, Stephen R. and Wilson, Edward O. (1993) *The Biophilia Hypothesis*. Washington, DC: Island Press.

Michalko, M. (1998) Thinking like a genius: eight strategies used by the supercreative, from Aristotle and Leonardo to Einstein and Edison. *Futurist*, 32: 21.

Orians, G. H. and Heerwagen, J. H. (1992) Evolved responses to landscapes. In Barkow, J., Tooby, J., and Cosmides, L. *The Adapted Mind*. Oxford: Oxford University Press.

Sternberg, E. M. (2009) *Healing Spaces: The science of place and well-being*. Cambridge, MA: Belknap Press.

Tuan, Y. (1968) *The Hydrologic Cycle and the Wisdom of God: a theme in geoteleology*. Toronto: University of Toronto.

Tuan, Y. (1974) *Topophilia: A study of environmental perception, attitudes and values*. Upper Saddle River, NJ: Prentice-Hall.

Wilson, E. O. (1984) *Biophilia*. Cambridge, MA: Harvard University Press.

Chapter 11

Green Pleasures[1]

Constance Classen

Sustainable Cities and the Senses

At first glance "sustainability" and "pleasure" seem at odds. "Green" practices are commonly thought to involve an almost puritanical restriction of
pleasures: shivering in frosty interiors to save on energy consumption,
forgoing exotic foods in favor of homegrown staples, or walking weary
miles to work rather than riding in comfort in a car. Surely "green" living
describes an ascetic rather than aesthetic lifestyle. Beyond the satisfaction
of feeling virtuous, what pleasures, what sensory enjoyments might living
in a sustainable city offer?

Potentially, a great many more than we currently enjoy. Modern
urban life, in fact, can be said to have deprived us of many traditional
sensory pleasures at the same time as it has inflicted on us many new
sensory displeasures. The early nineteenth-century utopian philosopher
Charles Fourier wrote of what he called the "sensory ills of civilization":
"The din of the trades . . . the sight of hanging rags, of the dirty dwellings
. . . the stifling smell of the drains . . . painfully affect the sight, hearing, and
smell."[2] Such unpleasant sensations had not been unknown to city
dwellers of earlier periods, but in the eighteenth and nineteenth centuries
they became greatly intensified by the immense growth of urban populations and the depletion of urban green spaces.

The increasing industrialization and mechanization of the city in
the twentieth century created a new set of urban sensations. Significantly,
while a list of noises compiled in nineteenth-century New York centered on
sounds produced by peddlers, street musicians, animals, and horse-drawn
vehicles, by 1925 the dominant noises of New York were said to come
from motorcars, subway trains, drills, and other mechanical sources.[3] It
seems that we must pay for the physical comforts and convenience
provided by modern technology with a good deal of sensory displeasure.
What is more, the sonic, chemical, and visual effluvia of the city are also
harmful to our physical well-being: noise pollution contributes to stress as

well as to hearing loss, smog harms our bodies and impairs our sense of smell, and excessive exposure to artificial light may contribute to a range of health problems.[4]

Whether deliberately planned as such or not, cities are inevitably "sensescapes" – landscapes of sounds and sights, smells and textures, and the flavors of its characteristic foods. As we rethink urban design within a context of ecological sustainability, we need to look for urban models that can fruitfully sustain our sensory lives. Indeed, perhaps the best way to encourage people to commit themselves to new modes of urban existence is by engaging them through pleasurable sensory experiences: green pleasures, rediscovered and reimagined within a revitalized cityscape.

Consider a scene common to many urban centers: a dingy street lined with uninviting buildings, jammed with cars, noisy with honking horns, reeking of exhaust, and offering little reason or opportunity for passersby to linger. Compare this scene with that presented by the Rua das Flores in Curitiba, Brazil. Rua das Flores was a typical traffic-congested downtown street until, in the 1970s, it was transformed into a pedestrian mall. The street's monotonous stretch of asphalt was replaced with cobblestones and enlivened by flowers, its historic buildings were renovated, and its cafés and kiosks were opened. Vehicular traffic was reduced within the city by an efficient, low-cost system of public transportation.

These measures have transformed urban life: where cars once dominated, flowers bloom, people relax on benches, birds are audible, street artists perform, and children play. Despite initial attempts by certain disaffected groups to reclaim the city core for cars, the public response has been enthusiastic. A "greener" street has turned out to be a more pleasurable street.[5]

While good sensory design is not always the result of conscious planning, it would be helpful to have certain guidelines when one is attempting to create an aesthetically pleasing sustainable city. I propose that the following serve as basic principles: first, that the widespread privileging of vision in modern urban life be tempered by an increased sensitivity to the nonvisual senses, to the "invisible city." Second, that an integrated diversity of sensory stimuli should generally be preferred to a tedious uniformity. Third, that the sensory design of a community be rooted in local cultural traditions and ecological systems. Fourth, that any program for the development of a green aesthetics be guided by an ideal of working in cooperation with nature and be grounded in social justice and compassion.

The Urban Sensescape

The city as sensescape has been the subject of many recent works and even one museum exhibition (Sense of the City, held at the Canadian Centre for Architecture in 2005).[6] A recurrent theme in studies of the sensory

Figure 11.1 The city of skyscrapers: downtown Toronto.

profile of modern cities is an emphasis on the visual. With few exceptions, sonic, tactile, and olfactory qualities are ignored in contemporary urban and architectural designs, while visual effects such as monumental height or striking appearance are celebrated. This is in keeping with the general rise in cultural importance of sight in modernity. Through longstanding cultural associations, sight has functioned as the sense of domination, detachment, display, and cleanliness (in contrast to the more "impure" sense of touch). These are all values highly esteemed in modernity and emphasized in the urban experience: the surveillance of well-lit city streets, the dominating and detached view from skyscrapers, the visual spectacle of the cityscape, the clean lines of modern buildings and paved streets.[7]

As has become obvious, however, these predominantly visual values have not served us well, as individuals, societies, or inhabitants of Earth. Domination leads to exploitation, detachment to disengagement, and conspicuous display to copious waste. Even the clean look of the modern city with its subterranean sewers, its electrical energy, its sleek buildings, its shiny cars, and its synthetic products has turned out to simply be a mask for immense waste. We are dismayed to learn that, in the end, dirt is "cleaner" than discarded plastic.

If this state of affairs has been partially brought about by overemphasis on the visual, how might we counter it through the elaboration of alternative sensory paradigms? In Fourier's utopian project, taste and touch were given priority over sight.[8] A taste-based paradigm for sustainable living has recently been proposed by the Slow Food Movement. According to this model, the promotion and sharing of local foods (as seen, for example, in the creation of the University of Gastronomic Sciences and an "Ark of Taste" to preserve traditional foods) provides the basis for recreating the city as a nurturing, small-scale, and cooperative environment – the Cittàslow.[9]

Members of the Slow Food Movement would likely approve of Fourier's celebration of "gastrosophy," or culinary wisdom, as the noblest science. Indeed, the movement's 2008 hosting of a pseudo "food Olympics" with flag-bearing representatives of countries around the world recalls Fourier's notion that in a utopian future international culinary competitions would replace political strife and wars.[10]

I am wary of this food-based model, attractive though it may seem. Critics of the Slow Food Movement have asserted that it promotes elitist standards of taste by characterizing certain foods as socially superior to others.[11] A more crucial concern, as I see it, is the distinction the movement draws between consumer and consumed, into which latter class potentially go all the animals and plants of the world. Once the natural world is perceived as food, it becomes symbolically dead. We see this in the language of Slow Food advocates when they use expressions such as "locally reared meat."[12] The distinction between consumers and consumed, perhaps inevitable in any taste-based social model, is not helpful in dealing with larger issues of environmental exploitation.

A full-bodied experience of the world requires all the senses. However, if we are to counter the domination of sight in contemporary culture, I suggest that we pay particular attention to touch. By cultivating tactile values of intimacy, interaction, and integration – values that promote engagement with our physical and social worlds – we can more effectively sustain both our cities and ourselves.[13]

The cleanest source of energy, some say, is muscle power,[14] and muscle power, by involving us in direct interaction with our physical surroundings, provides us with one of our greatest sources of pleasure. Labor-saving devices, from leaf blowers to cars, while seeming to make life easier, in fact impoverish us by diminishing our range of physical experiences and disengaging us from our environments. As E.F. Schumacher noted in his classic, *Small is Beautiful*, "The type of work which modern technology is most successful at reducing or even eliminating is skilful [sic.], productive work of human hands, in touch with real materials of one kind or another."[15] A tactile city would offer opportunities for citizens to engage with "real materials," not only through the work of the hand, such as gardening or craftwork, but through maximizing the possibilities for and pleasures of walking, as well as other physical activities. A tactile city would also aim to increase opportunities for social interaction, such as the participation of the public in communal events or the informal encounters that occur on pedestrian streets like Curitiba's Rua das Flores.

Nonetheless, when I propose a move away from visual values to tactile values, I am not saying that our cities are already so visually appealing that no more trouble need be taken on that account. I am suggesting that we need to seek out smaller, more intimate beauties rather than grand visual effects (like those of monuments and skyscrapers) if we want to create a more pleasurable multi-sensory environment.[16]

　　　　Were there "tactile cities" in the past? One could argue that the medieval city, with its richly textured buildings; narrow, winding streets; and intricate craftwork offered as much for touch as for sight. Medievals accorded great value to touch as the sense that provided reliable information about the world, compared with sight, which could be readily deceived

Figure 11.2 Medieval textures: cobble stones in Oxford.

by surface impressions.[17] A keen sense of touch was associated with mental acuity. Thomas Aquinas declared: "Among men it is in virtue of fineness of touch, and not of any other sense, that we discriminate the mentally gifted."[18]

Manual skills were highly cultivated in the Middle Ages, whether on the small scale of weaving or the large scale of church building. As Lewis Mumford noted, the medieval builder was a man who knew his materials, his tools, and his workers, as compared the more visually oriented architect of later days, who knew his texts and his blueprints.[19] The visual impact of their great cathedrals notwithstanding, medievals did not manifest much interest in grand views. Their stories, paintings, and carvings – even when they deal with religious subjects – typically depict what is close at hand: family and work, domestic animals, flowers, and fruits. In the fourteenth-century Canterbury Tales, though the pilgrims are on the road, we get no vistas or landscapes. Rather, we are drawn into intimate bodily experiences, such as "kneling . . . upon the small and soft and swete gras".[20] Medieval social life was itself highly tactile with its communal forms of domesticity, labor, and worship. Whether at work or at prayer, while eating or sleeping, close proximity with others was the norm.[21] While exploring the tactile values of the Middle Ages will not provide us with an aesthetic model for contemporary culture, it may serve to stimulate our sensory imaginations and help us conceive alternatives to our visual obsessions.[22]

Sensory Diversity

One characteristic of urban life that is both unsustainable and non-pleasurable is the unnatural uniformity of light and temperature levels in many residences and public buildings. The American Society of Heating, Refrigerating, and Air-Conditioning Engineers has determined standards of thermal comfort employed across the United States and consulted by other countries. Yet numerous studies have shown that people living in different cultures and climates have different thermal comfort zones; no one standard will suit all.[23] Furthermore, as Forrest Wilson notes in his analysis of the role of perception in design, research approaches to the thermal environment "concentrate on preventing feelings of discomfort rather than on producing positive responses".[24] While thermally neutral environments do not distract us, they also do not stimulate us.

To have the same temperature and the same lighting every-where, every day, is akin to being served a meal of one taste every day. Temperature and light are most pleasurable when they provide a diversity of sensations, as do the warmth of a fire on a cold day, a cool garden in the heat of summer, a ray of sunshine in a dusky interior. In his In Praise of Shadows, Jun'ichiro Tanizaki contrasts the Japanese appreciation of the subtle variations of light and darkness with the Western quest for total brightness.[25] In Thermal Delight in Architecture, Lisa Heschong explores

Figure 11.3 Kinaesthetic pleasure: a footpath transformed into a skating path in Magog, Quebec.

the range of thermal strategies that has been employed across cultures and in history, from the medieval use of tapestries for insulating warmth to the Middle Eastern creation of courtyards as cool interior spaces.[26]

Like wood and stone, temperature and light can be crafted to provide a more stimulating environment with a smaller expenditure of energy. Not all areas of a house, for example, need be equally bright or equally warm, since different areas have different uses. A reading nook would preferably be sunny and warm, a bedroom cool and dim, its windows perhaps shaded by trees. Having one particularly cool room in the summer or one especially warm room in the winter will furthermore serve as a social magnet, bringing people in the household together. Anthropologist Lawrence Wylie gives an instance of this when he describes how his family grew more intimate after moving from the United States, "where a movement of a finger regulates the heat of the whole house," to a house in a French village, where "the fire of oak logs which burned day and night for six months became the focal point of our family life."[27] Also persuasive is the argument that the sense of pleasure or comfort is increased by a preceding sense of displeasure or discomfort: the warmth of a living room entices because of the coolness of a bathroom; a balmy spring day is especially appealing after a long, cold winter.

What holds for light and temperature also holds true for the other sensory stimuli that could and should be taken into account by urban designers.[28] Lilacs, for example, may offer a delightful fragrance and a pleasing greenery, but to encounter them on every city street would be monotonous. There is no blanket sensory solution for the sensory displeasures of the city, which themselves blanket us with malodors, noises, and glaring lights. A diversity of sensory stimuli is necessary, an integrated diversity that seeks to promote a sense of coherence without jarring effects (e.g., living in Spanish-style villas, designed for a warm climate in cold Northern cities only magnifies discomforts).[29] Jane Jacobs' Hudson Street in Greenwich Village in the late 1950s provided such sensuous diversity mainly because of an abundance of small shops, cafés, restaurants, and the like.[30]

The hand of the urban planner should not be felt everywhere, however, for this too would create monotony. One should not have to ask, "Is there no tree in this city which has not been specifically planted to provide us with the correct amount of shade and the recommended dose of fragrance?" As every child who prefers an overgrown, vacant lot with its wildflowers to a carefully planned playground knows, the city needs wild spaces.

The Local Touch

Although we now live in a global village, we hardly wish to encounter the same village, no matter how charming, everywhere we go. To be attractive, meaningful, and sustainable, urban design needs to be grounded in local environments and traditions. This, of course, runs counter to the modern trend for cities everywhere to look, feel, and smell alike. Medievals had to live on local produce and build with local materials. These restrictions, however, provided them with a satisfying sense of place, something we can still sense when we see a European village that looks as though it has grown out of a hillside. Our increasing awareness of the high environmental and financial costs of transporting goods and materials over long distances may well lead us to rediscover the pleasure, as well as the utility, of engaging with local materials and making the best out of what we have at hand.

Urban design, likewise, is more meaningful and pleasing if it takes into account the cultural traditions that we have "on hand." Every society has its own sensory preferences and customs. In my essay "McLuhan in the Rainforest," I explored the diverse "sensory models" of different indigenous societies. The Tzotzil of Mexico, who associate heat with the life force, give particular meaning to the thermal values of their environment: the coolness of the earth, the warmth of the sun, the differing climates of the highlands and the lowlands. In contrast, the Ongee of the Andaman Islands know the world as a "smellscape" and pay close attention to the different odors of humans, animals, and plants.[31] A culturally sensitive urban design will create an aesthetic environment that is not just "nice" to look at or "nice" to smell but that is meaningful for its inhabitants because it is pervaded by local traditions and sensory values.

This perceived meaningfulness of sensory experiences will produce the deepest pleasures.

Our Fair City

A city may offer a range of pleasurable sensory experiences and yet be rife with social problems that prevent people from enjoying these pleasures. The hanging gardens of ancient Babylon must have been delightful to see and smell but much less so for Babylon's countless slaves. Similarly, ecologically helpful rooftop gardens and green roofs in contemporary cities would gratify many but certainly not the homeless below. A "fair" city is not necessarily a fair city.

Furthermore, while I have emphasized the importance of grounding the sensescape of the city in local culture, this should not be interpreted as promoting an uncritical celebration of traditional practices or products.

A culture's culinary heritage, for example, may include practices that are environmentally unsound, such as cutting down endangered palm trees to extract their edible hearts, or cruel, such as force-feeding geese to produce *pâté de foie gras*. A green aesthetics requires a radically new way of thinking about our relationship to Earth, and while it can and should seek to make connections with local traditions, it cannot recreate some mythical past.

If a return to a preindustrial past is no answer, neither is the mechanical model of environmental domination that produced the current

Figure 11.4 An urban oasis: a rooftop garden in London.

crisis. According to this model if we could just engineer pigs and cars to produce fewer environmentally harmful emissions and induce worms and microorganisms to recycle our waste products, we could keep our factory farms, superhighways, and big-box stores and go on as before.[32] But even if technological ingenuity could achieve these ends, it cannot lessen our alienation from nature and the deprivations and discords this produces in ourselves and the world. The pleasure of walking through a forest cannot be bought in a store, and forests lose their viability as wildlife habitats when they are bisected by highways and hedged in by suburban developments.[33]

The aesthetic of sustainability is not about recovering preindustrial ways of life or making cities into green machines for living. Rather, such an aesthetic calls for new ways of perceiving and interacting with Earth and its inhabitants based on justice, compassion, and cooperation – the sharing of pleasure. It would help if we thought of green pleasures not just as green insofar as they promote sustainable practices, but also insofar as they cultivate a more ecological way of relating to the world with both our minds and our bodies.

Notes

1. Originally published in *Harvard Design Magazine*'s Fall/Winter 2009–2010 edition.
2. Cited in Constance Classen, *The Color of Angels: Cosmology, Gender and the Aesthetic Imagination* (London: Routledge, 1998), 27.
3. Emily Thompson, "Noise and Noise Abatement in the Modern City," *Sense of the City: An Alternative Approach to Urbanism*, ed. Mirko Zardini (Montreal: Canadian Centre for Architecture and Lars Müller Publishers, 2005), 190–191.
4. See, e.g., Roger Highfield, "Noise Having Huge Impact on Health," *Daily Telegraph*, August 23, 2007, www.telegraph.co.uk/news/1561091/"Noise-having-huge-impact-on-health.html (accessed February 18, 2013); Mica Rosenberg, "Mexico City Smog Hurting People's Sense of Smell," Reuters, June 4, 2008, www.reuters.com/article/environment-News/idUSN0430376020080604 (accessed February 18, 2013); and Ron Chepesiuk, "Missing the Dark: Health Effects of Light Pollution," *Environmental Health Perspectives*, January 2009, www.ehponline.org/docs/2009/117–1/focus-abs.html (accessed February 18, 2013).
5. For an overview of Curitiba as a model of a sustainable city, see Bill McKibben, "Curitiba and Hope," *Mother Jones*, November 8, 2005, www.motherjones.com/environment/2005/11/curitiba-and-hope (accessed February 18, 2013).
6. See, e.g., "The Senses and the City," special issue of *The Senses & Society*, ed. Mags Adams and Simon Guy, July 2007; *The City and the Senses: Urban Culture Since 1500*, ed. Alexander Cowan and Jill Steward (Aldershot: Ashgate, 2007); Mirko Zardini (ed.), *Sense of the City: An Alternative Approach to Urbanism* (Montreal: Canadian Centre for Architecture and Lars Müller Publishers, 2005); John Urry, "City Life and the Senses," in *A Companion to the City*, ed. Gary Bridge and Sophie Watson (Oxford: Blackwell Publishers, 2000), 388–397.
7. Two key works discussing the cultural role of vision in modernity are Guy Debord, *The Society of the Spectacle*, trans. Donald Nicholson-Smith (New York: Zone Books, 1967); and Martin Jay, "Scopic Regimes of Modernity," in *Vision and Visuality*, ed. Hal Foster

(Seattle: Bay Press, 1988), 3–38. I examine the increasing importance given to visual values in modernity in *Worlds of Sense: Exploring the Senses in History and across Cultures* (London: Routledge, 1993).

8. Classen, op. cit., 28.

9. Geoff Andrews, *The Slow Food Story: Politics and Pleasure* (Montreal: McGill-Queen's University Press, 2008); and Sarah Pink, "Sense and Sustainability: The Case of the Slow City Movement," Local Environment, March 2008, 95–106. For information on the Ark of Taste, see the Slow Food Foundation for Biodiversity, www.slowfoodfoundation.org/eng/fondazione.lasso (accessed February 18, 2013).

10. Classen, op. cit., 29.

11. Andrews, op. cit., 172.

12. Pink, op. cit., 103.

13. For an exploration of the social importance of touch, see *The Book of Touch*, ed. Constance Classen (Oxford: Berg, 2005).

14. Tracy Bhamra and Vicky Lofthouse, *Design for Sustainability* (Aldershot: Gower, 2007), 46.

15. E.F. Schumacher, *Small is Beautiful: Economics as if People Mattered* (New York: Harper & Row, 1973), 141.

16. The importance of touch in urban design and in architecture is discussed in David Howes, "Skinscapes: Embodiment, Culture and Environment," in *The Book of Touch*, 27–39; Joy Monice Malnar and Frank Vodvarka, *Sensory Design* (Minneapolis: University of Minnesota Press, 2004), 144–52; Juhani Pallasmaa, "Hapticity and Time," *Architectural Review*, May 2000, 78–84; and Richard Sennett, "The Sense of Touch," *Architectural Design*, March/April 1998, 18–23.

17. Robert Mandrou writes that premodernity touch "verified perception, giving solidity to the impressions produced by the other senses, which were not as reliable." Introduction to *Modern France, 1500–1640: An Essay in Historical Psychology*, trans. R.E. Hallmark (New York: Holmes & Meier, 1975), 53.

18. Thomas Aquinas, *Commentary on Aristotle's 'de Anima'*, trans. Kenelm Foster and Silvester Humphries (Chicago: St. Augustine's Press, 1995), 152.

19. Lewis Mumford, *Sticks and Stones: A Study of American Architecture and Civilization*, second edition (New York: Dover, 1955), 164.

20. Cited in Yi-Fu Tuan, *Passing Strange and Wonderful: Aesthetics, Nature, and Culture* (New York: Kodansha, 1995), 136.

21. Norbert Elias discusses the social importance of physical contact in the Middle Ages in *The Civilizing Process*, ed. Eric Dunning, Johan Goudsblom, and Stephen Mennell, trans. Edmund Jephcott (Oxford: Blackwell Publishers, 1994).

22. For a discussion of the environmental sustainability of the medieval city, see Lewis Mumford, *The Culture of Cities* (New York: Harcourt, Brace and Company, 1938), 42–51.

23. The considerable literature on this subject across disciplines includes Jie Han, Guoqiang Zhang, Quan Zhang, Jinwen Zhang, Jianlong Liu, Liwei Tian, Cong Zheng, Junhong Hao, Jianping Lin, Yanhui Liu and Demetrios J. Moschandreas, "Field Study on Occupants' Thermal Comfort and Residential Thermal Environment in a Hot-Humid Climate of China," *Building and Environment*, December 2007, 4043–4050; Igor Knez and Sofia Thorsson, "Influences of Culture and Environmental Attitude on Thermal, Emotional and Perceptual Evaluations of a Public Square," *International Journal of Biometeorology*, May 2006, 258–268; R.M.A. Humphreys, "Thermal Comfort Temperatures Worldwide: The Current Position," *Renewable Energy*, 8, 1–4 (1998), 139–44; Lisa Heschong, *Thermal Delight in Architecture* (Cambridge, MA: MIT Press, 1990), 16.

24. Cited in Malnar and Vodvarka, *Sensory Design*, 268.

25. New Haven, CT: Leete's Island Books, 1977.

26. Heschong. See also John S. Reynolds, *Courtyards: Aesthetic, Social, and Thermal Delight*, 1990 (New York: John Wiley & Sons, 2002).

27. Lawrence Wylie, *Village in the Vaucluse*, second edition (Cambridge: Harvard University Press, 1965), 145–146.

28. In *Sensory Design*, Malnar and Vodvarka discuss ways in which the built environment would benefit from multisensory design.

29. Norman Pressman criticizes the prevalence of architecture suited to warm climates in Northern cities in "The Idea of Winterness: Embracing Ice and Snow," in *Sense of the City*, ed. Mirko Zardini (Montreal: Canadian Centre for Architecture and Lars Müller Publishers, 2005), 129–41.

30. Jane Jacobs, *The Death and Life of Great American Cities* (New York: Vintage Books, 1961).

31. Constance Classen, "McLuhan in the Rainforest: The Sensory Worlds of Oral Cultures," in *Empire of the Senses: The Sensual Culture Reader*, ed. David Howes (Oxford: Berg, 2005), Chapter 8.

32. For examples of this approach, see C.W. Forsberg *et al.*, "The Enviropig Physiology, Performance, and Contribution to Nutrient Management, Advances in a Regulated Environment: The Leading Edge of Change in the Pork Industry," *American Journal of Animal Science* 81 (2003), www.animal-science.org/content/81/14_suppl_2/E68.full (accessed February 18, 2013); Mary Appelhof, *Worms Eat My Garbage: How to Setup and Maintain a Vermicomposting System* (Kalamazoo, MI: Flower Press, 1997); and Kenneth N. Timmis, Robert J. Steffan, and Randel Unterman, "Designing Microorganisms for the Treatment of Toxic Wastes," *Annual Review of Microbiology*, October 1994, 525–557.

33. Wilfried Wang examines the social underpinnings of environmental pollution in "Sustainability is a Cultural Problem," *Harvard Design Magazine*, Spring/Summer 2003, 1–5.

Index

Please note that page numbers relating to Notes have the letter 'n' following the page number. References to Figures are in *italics*.

Index